SO-AQM-727

Acknowledgements

To my parents, Ferd & Rita Niemann

To the newspaper editors who carry the *Invention Mysteries* syndicated column in their newspapers each week

To you, the reader.

About the author

Paul's interest in inventions goes all the way back to his teenage years. He was fascinated by the thought of inventing a product and having people want to purchase it. There were two inventions that inspired him ... the board game *Trivial Pursuit* and Spandex shorts. Believing that he was not the inventor type, though, he went to work for an ad agency in 1992 in St. Louis after receiving his M.B.A. degree. His first big break came later that year when he was assigned to launch a new product onto the market for the agency. After that, he started his own company, Market Launchers, with the purpose of helping inventors bring their inventions to market.

In 1998, he finally invented something of his own ... the IMPEACHMENT Card Game. The IMPEACHMENT game received nationwide media coverage and, soon after that, Paul began writing a regular column for the magazine for inventors, *Inventors' Digest*. It was while writing for *Inventors' Digest* in 2002 that he came up with the idea of creating his own weekly newspaper column. This column became *Invention Mysteries*, and this book is the collection of the first 47 stories that appeared in newspapers during 2003.

Paul also teaches marketing at Quincy (IL) University, and he writes *Invention Mysteries* from the horse farm on which he grew up. This is where he came up with the idea for the name of his publishing company, Horsefeathers.

Thanks, and enjoy "the little-known stories behind well-known inventions!" We'll do it again next year.

What others are saying about Invention Mysteries

"Paul Niemann's Invention Mysteries column is a perfect blend of drama, enlightenment, and just plain fun, appealing to readers of all ages."
– J. Simms, Mexico, MO

"... have enjoyed the many informational and neat-to-know items and stories."
– B. Weir, Clay Center, KS

"Just wanted to drop you a quick note and tell you how much I enjoy your Invention Mysteries column. I wish eveyone in America, and the world, could read what you have written."
– B. Atkins, Jonesville, LA

"Enjoyed your article on Da Vinci the inventor"
– B. Boles, via e-mail

"Keep Invention Mysteries coming. Quite informative and interesting. I have learned things about inventors and inventions that I had not been exposed to before."
–H. Helm, Comanche, TX

"... the articles are entertaining and well written. Most enjoyable!"
– G. Nelson, via e-mail

"Good job. You have a style that is readable and factual."
– R. Scheinkman, St. Louis, MO

"Paul Niemann's insightful look into the mysteries of many common and not-so-common inventions, and the people behind them, adds a dimension of variety."
– D. Osborn, Ozark, MO

INVENTION MYSTERIES ...
... the little-known stories behind
well-known inventions

Paul Niemann

Published by:

Horsefeathers Publishing
P.O. Box 5148
Quincy, IL 62305
www.InventionMysteries.com

All rights reserved. No part of this book may be reproduced or transmitted in any form without written permission from the author.

Copyright © 2004 Paul Niemann
Printed in the United States of America
Illustrations by Kevin Cordtz
Book layout and typesetting by Randy McElhoe

Library of Congress Control Number: 2003115777

Niemann, Paul
Invention Mysteries: the little-known stories behind
well-known inventions / Paul Niemann. - 1st ed.
P.cm.
Includes index.
ISBN 0-9748041-0
1. Inventions
2. Inventors
3. Little-known stories
4. Mysteries

2003-115777

Table of Contents

*"Inventors are the creators
of the world – after God"*
Mark Twain

Here's why you've never heard of the other person who invented the telephone

We all know that Alexander Graham Bell is credited with inventing the telephone, but did you know that there was another person who tried to patent a different version of the telephone on the very same day as Bell in 1876?

Born in Ohio in 1835, he was a physics professor at nearby Oberlein College, and was a renowned inventor due to the musical telegraph that he invented. Little is known about him because, in what has to be one of the worst cases of being "a day late and a dollar short," he arrived at the patent office two hours after Bell arrived to apply for a patent for his version of the telephone.

His name is Elisha Gray and, as a result of arriving two hours after Bell arrived, most of the world has never heard of him.

What happened?

U.S. patent law states that the first one to invent a new product is the rightful owner of the product, regardless of who applies for a patent first. Adequate records are necessary whenever there is a dispute. Since Bell applied for his patent first, he was initially awarded the patent.

Gray did prevent the issuance of Bell's patent temporarily, however, pending a legal hearing. Since he did not keep adequate records of his design, however, he lost any possible rights as Bell's right to the patent was later sustained by the U.S. Supreme Court and the rest, as they say, is history.

The basis of Gray's legal action against Bell was that Bell had filed for his patent before he had a working model of his telephone, according to Inventors' Digest magazine. But the Supreme Court ruled that a person can prove that his invention is complete and ready for patenting even before a working model has been produced, a ruling that later served as a precedent on a similar type of lawsuit years later.

Gray was not the only other person to stake a claim to inventing the telephone. Daniel Drawbaugh, who was born near Harrisburg, Pennsylvania, claimed to have invented the telephone long before Bell filed a patent application in 1875. Drawbaugh didn't have any papers or records to prove his claim, though, and the Supreme Court rejected his claims by four votes to three. Alexander Graham Bell, on the other hand, had kept excellent records.

Elisha Gray did go on to invent other products, such as the facsimile telegraph system that he patented in 1888. Bell, who was born in Edinburgh, Scotland in 1847, became a U.S. citizen in 1882. He went on to become one of the co-founders of the National Geographic Society, and he served as its president from 1896 to 1904.

Elisha Gray, however, has been forgotten by much of the world.

Was Bell's telephone greeted with enthusiasm by everyone at the time?

As is the case with many new inventions, there were those who rejected the telephone for one reason or another. Even President Rutherford B. Hayes was skeptical of the new device when Bell demonstrated it to him at the White House in 1876.

There was also a well-known "investor" who had an opportunity to invest in the telephone directly with Bell, but he rejected the opportunity. According to his writings, he was a big fan of new inventions, but since he had previously invested in several that had failed, he turned down a chance to invest in the telephone. Who was he?

Mark Twain, who patented two of his own inventions.

Which U.S. Presidents were the most successful inventors?

Since my hometown of Quincy, IL is named after a U.S. President -- John Quincy Adams, our 6th President -- I decided to use this column to focus on Presidents who toiled as inventors. By the way, there are 12 states that contain a Quincy: California, Florida, Illinois, Indiana, Kansas, Kentucky, Massachusetts, Michigan, Missouri, Ohio, Pennsylvania and Washington.

While Thomas Jefferson, our nation's 3rd President, was the most accomplished inventor among all the U.S. Presidents, he did not hold a patent on any of his inventions. Only one President ever received a patent, and only one received a trademark. Who were they? Read on; the answers are at the end of the column.

Among Thomas Jefferson's inventions were such devices as a macaroni machine that he invented in 1787, the swivel chair, the spherical sundial, the moldboard plow and the cipher wheel, which was an ingenious way to allow people to code and decode messages. Jefferson's cipher wheel was used until 1802, and then it was "re-invented" just prior to World War I and used by the US army and other military services to send messages back and forth. Jefferson served as American minister to France in the 1780's and, as a result of his travels throughout Europe, was able to adapt some of the things he saw in Europe to benefit Americans as well.

Jefferson felt that all people should have access to new technology and, since he didn't want others to be deprived of the benefits that new inventions bring, he never applied for a patent on any of his inventions. He considered patents to be an unfair monopoly.

Several of Thomas Jefferson's inventions are still in use today; they deal mainly with agricultural and mechanical products.

He also was responsible for introducing French fries into the United States.

One of Jefferson's most notable achievements was the founding of the University of Virginia, and this was one of only three achievements that he had listed on his tombstone.

Jefferson's impact on the United States patent system can be seen today in the fact that each new patent application must meet three criteria before being issued a patent. A patent must be: New, not obvious, and useful. While Jefferson was the most prolific of any presidential inventor, he wasn't the only President to have some success at inventing.

In two separate boating incidents, one as a teenager on the Mississippi River and one on the Great Lakes, President Lincoln got his boats stuck in shallow waters, known as "shoals." These two experiences inspired Lincoln to invent a solution to help him navigate his boat through shallow waters. A wooden model of this invention, which Lincoln made himself, is in the Smithsonian Institution. The invention was never sold to the public, though.

In 1858, Lincoln called the introduction of patent laws one of the three most important developments "in the world's history," along with the discovery of America and the perfection of printing.

During the Civil War, he took a personal interest in the development of new types of weapons: iron ships, the observation balloon, the breech-loading rifle and the machine gun.

President Washington was also a successful inventor, and in 1772 he received a trademark for his brand of flour.

Since there haven't been very many Presidents who were considered inventors, I guess you could fish for additional Presidential inventions by insisting that President Nixon invented impeachment, and that President Clinton holds the current patent on it.

While Thomas Jefferson invented the most new products of all the Presidents, only one U.S. President has ever received a patent, and it wasn't Jefferson. Do you know which President received a patent?

A. George Washington
B. Abraham Lincoln
C. Teddy Roosevelt
D. Harry Truman

ANSWER: President Lincoln was issued Patent # 6,469 for "A Device for Buoying Vessels Over Shoals" on May 22, 1849 while still a Congressman in Illinois. If you guessed George Washington, you were close; he is the only President to receive a trademark, which he received in 1772 for his brand of flour.

Who invented the World Wide Web – and why did he choose to NOT patent it?
(HINT: It wasn't Al Gore)

"Oh, I see they've got the Internet on computers now."

Which famous "person" said this?

If you guessed that it was former vice president Al Gore, you were close. While Mr. Gore didn't actually make the above statement, he did claim to play a major role in creating the Internet and, as a result, has often been taken to task for his claim.

The sole inventor of the World Wide Web is Timothy Berners-Lee, an English computer scientist. The birth of his brainchild began in 1989, when he envisioned a way to link documents on the Internet so that web surfers could jump from one document to another through hyperlinks.

The Web has revolutionized the computer world, as there were more than 140 million Americans online in 2001 -- which translates into 54 percent of the U.S. population -- and the number of Americans using the Web continues to grow each day. It almost seems that nearly every man, woman and child now has their own web site.

Many people don't realize that there is a big difference between the Internet and the World Wide Web, since the two terms are often used interchangeably.

The Inter**net** is a **net**work of computers all linked together through phone lines, while the Web is connected by hypertext links. It is the Web that people are talking about when they refer to the "Information Superhighway."

The Internet can exist without the Web, but the Web could not exist without the Internet. The Web made the Internet useful because people are interested in information rather than a system of computers and cables.

Berners-Lee has been hailed by *Time* magazine as one of the 100 greatest minds of the 20th century. His invention has greatly changed the way people do business, as millions of Americans now use the Web for all kinds of purchases, research and other functions. Time even suggested that the Web may be as important an invention as Gutenberg's printing press, which was developed in the mid-1400's.

Berners-Lee could have patented the World Wide Web and made money each time someone visits a web site, but instead his desire was for everyone to be able to benefit from the Web, and this is the reason why he chose to NOT patent it. He passed up a fortune so that the world would benefit from it.

Another reason for not patenting his technology was because he feared that the Web wouldn't develop as fully if he patented it. He felt that it was necessary to make the Web an open system in order for it to be universal and allow others to contribute to its development, and he has fought to keep it open and free.

The full impact of his invention has yet to be fully known; Berners-Lee thinks that the Web will eventually be able to reason with humans.

What is the inventor of the World Wide Web doing now?

Berners-Lee is currently the director of the World Wide Web Consortium, the coordinating body for Web development, and he also occupies the 3Com Founders chair at the MIT Laboratory for Computer Science.

So who was it that made that infamous remark at the beginning of this column, *"Oh, I see they've got the Internet on computers now?"*

None other than Homer Simpson. D'oh!

"Everything that can be invented – has already been invented," and other famous invention-related quotes

This (in)famous quote has long been attributed to the commissioner of the United States Patent Office, a Mr. Charles Duell, back in 1899, and it's one of my personal favorites. Since the patent office opened in 1790, it has registered more than 6.2 million patents.

When I called the patent office in an attempt to verify the accuracy of this quote, the lady on the other end of Mr. Bell's invention told me that the commissioner was misquoted when he said that. Whether or not she was being honest is anyone's guess; you really couldn't blame her if she was simply towing the company line in order to avoid embarrassment.

Either way, the jury is still out on whether the former patent commissioner is an early inventor of the art of being misquoted.

Surprisingly, though, this wasn't the first time someone had made such a remark. In 10 A.D., Roman Engineer Julius Sextus Frontinus said, *"Inventions have long since reached their limit, and I see no hope for further developments."*

This began a long line of infamous invention-related notable quotables, such as:

"That's an amazing invention, but who would ever want to use one of them?" ... President Rutherford B. Hayes in 1876, after Alexander Graham Bell demonstrated the telephone to him at the White House.

"There is no likelihood man can ever tap the power of the atom," ... Robert Milken, Nobel Prize winner in physics, 1923

"Heavier-than-air flying machines are impossible,"
– Lord Kelvin, President Royal Society, 1895

"Who the hell wants to watch movies with sound?" Who said this? Believe it or not, it was the president of Warner Brothers Studios, Harry Warner, sometime around 1918.

... and one of the all-time favorite infamous baseball quotes ...

"Ruth made a grave mistake when he gave up pitching. Working once a week, he might have lasted a long time and become a great star" ... Tris Speaker on Babe Ruth's future, 1921.

While this quote doesn't deal directly with inventions, it does relate to a new idea – putting Babe Ruth in right field full-time. The possibility of the Babe once being a pitcher sounds pretty crazy, doesn't it?

About as crazy as the former patent commissioner saying that, *"Everything that can be invented – has already been invented."*

While no one wants to claim credit for such ridiculous quotes as those mentioned above, there are some memorable quotes worth claiming. Do you know who said each of the following quotes?

1. "Necessity is the mother of invention."
2. The patent system "added the fuel of interest to the fire of genius, in the discovery and production of new and useful things." Hint: He was a U.S. President who did some inventing of his own.
3. "Anything that won't sell, I don't want to invent. Its sale is proof of utility, and utility is success."
4. "A country without a patent office and good patent laws ... couldn't travel any way but sideways or backwards."
5. "People think of the inventor as a screwball, but no one ever asks the inventor what he thinks of other people."

ANSWERS:
1. Ralph Waldo Emerson is often credited with this quote, but the 17th and 18th century playwrights William Wycherley and Richard Sheridan both made variations on this phrase. Emerson actually said, "invention breeds invention."
2. Abraham Lincoln
3. Thomas Edison
4. Mark Twain
5. Charles Kettering

Screen Doors for Submarines and Lead Balloons:
21 of the craziest inventions of all time

In the previous story, we brought you some of the most interesting and odd invention-related quotes in the article entitled,
"Everything that can be invented – has already been invented."

Now you'll see some odd inventions that were just not meant
to be. To get the maximum amount of enjoyment out of this
article, try to visualize what each of these must-see, must-have
inventions might have looked like. By the way, that quote
about everything already being invented has long been attributed to the former head of the U.S. patent office in 1899.

Can you guess which of these inventions were actually patented, and which ones were left on the cutting room floor -- send-
ing the inventor back to the old drawing board? The answer
appears at the end of the column.

1. Flying saucer submarine
2. A method of growing unicorns
3. A dog watch. I presume this is for the busy executive dog
    on a tight schedule? And does every hour equal 7 hours in
    dog hours?
4. An amphibious horse drawn light vehicle, which is used by
    a horse walking in shallow water. Does it come with a
    water bucket in case the horse gets thirsty?
5. A leash for walking an imaginary dog, which produces a
    variety of barks, growls, etc. A similar version of this
    actually made it onto the market back in the '70's!
6. Toilet landing lights
7. A Santa Claus detector, which signals the arrival of
    Santa Claus. If this one really exists, would there be a
    debate over whether or not it really exists?
8. A method of creating an anti-gravity illusion

9. A drive-thru ATM machine with instructions written in Braille (think about it)
10. A device for producing dimples. And you thought people were just born that way!
11. A haircut machine that sucks in your hair like a vacuum cleaner, and then gives it a perfect cut
12. A motorized ice cream cone. Don't you wish you would have had *that* as a kid?
13. A drip pan for caskets (in case the dead leak!)
14. A jet-powered surfboard
15. An all-terrain baby stroller. For the adventuresome ittle tykes!
16. A pet petter. This device has a human-like hand that pets Rover when you're not able to.
17. A slingshot golfing system. This device slings the little white ball, then converts into a putter once you reach the putting green.
18. A human slingshot machine
19. A gas-powered snow ski fan. For those who live in the Midwest and other mountainless areas.

Check out these nifty little inventions from across the pond: Americans don't have a monopoly on ridiculous patents, so we present you with two of Great Britain's craziest inventions:

20. A horse-powered minibus, in which the horse walks along a treadmill in the middle of the bus to drive the wheels via a gearbox.
21. A ladder which enables spiders to climb out of the bathtub

So which of these 21 "inventions" were actually patented, and which ones were left on the cutting room floor?

ANSWER: All of the above inventions were patented! And who was it that created # 8 -- a method of creating an anti-gravity illusion?

Michael Jackson. Yes, *that* Michael Jackson, is listed on the patent as a co-inventor. The patent explains that the invention allows a person to "lean forward beyond his center of gravity by ... wearing a specially designed pair of shoes." This, of course, refers to the moonwalk that Michael Jackson made famous.

Sources: www.Patent.freeserve.co.uk,
www.TotallyAbsurdInventions.com and
http://inventors.about.com

We might have lost World War II if not for this little-known "invention"

In war, the side with the superior level of intelligence has a major advantage, as does the side with the most effective use of communications.

This story delves into how the use of a certain communications tool influenced World War II. This "invention," though not patentable, is probably more native to America than apple pie and baseball.

It was used in every assault the U.S. Marines conducted in the Pacific from 1942 to 1945 -- Guadalcanal, Tarawa, Peleliu and Iwo Jima. It was used in all six Marine divisions, Marine Raider battalions and Marine parachute units, enabling our men to transmit messages by telephone and radio in a code the Japanese never broke.

Without it, the Marines would never have taken Iwo Jima, according to Major Howard Connor of the 5th Marine Division. So what is this great "invention" that helped us win World War II?

Navaho code talkers. These code talkers were Navaho Indians who were recruited to transmit and interpret messages during the war.

The Germans had the Enigma machine as their code system, but it was no match for the Navajo code talkers. Its syntax and tonal qualities, not to mention dialects, make it unintelligible to anyone without extensive exposure and training. It has no alphabet or symbols, and is spoken only on the Navajo lands of the American Southwest. One estimate indicates that less than 30 non-Navajos, none of them Japanese, could understand the language at the outbreak of World War II.

How important to the American war effort were the

Navaho code talkers?

Major Connor had six Navajo code talkers working around the clock during the first two days of the battle of Iwo Jima. Those six sent and received over 800 messages, all without error.

The subject of a 2002 Hollywood movie called *Windtalkers*, the Navaho code talkers' code is the only unbroken code in modern military history. The code enabled American translators stationed outside the United States to decipher the code in minutes, whereas other codes would take approximately two hours to decipher. It would take only 20 seconds for the Navaho code talkers to decode a three-line English message, whereas machines required 30 minutes to perform the same job.

So how did the Navaho code talkers go virtually unnoticed for half a century after the war had ended?

Navajo remained potentially valuable as code even after the war. For that reason, the code talkers, whose skill and courage saved both American lives and military engagements, only recently earned recognition from the U.S. government and the public. They were honored in September of 1992 at the Pentagon; the Navajo code talker exhibit is a regular stop on the Pentagon tour. The exhibit includes a display of photographs, equipment and the original code, along with an explanation of how the code worked.

Thirty-five Navajo code talkers, all veterans of the U.S. Marine Corps, and their families traveled from their homes on the Navajo Reservation, which includes parts of Arizona, New Mexico and Utah, to attend the dedication of the Navajo code talker exhibit.

Who was this visionary Navaho individual who came up with this brilliant plan to use their native language as code in World War II?

The idea actually came from an American named Philip Johnston, the son of a missionary to the Navajos and one of the few non-Navajos who spoke their language fluently. Johnston, reared on the Navajo reservation, was a World War I veteran who knew of the military's search for a code that would withstand all attempts to decipher it. He also knew that Native American languages -- notably Choctaw -- had been used in World War I to encode messages.

Sources: The History Channel, Naval Historical Center of the Dept. of the Navy.

What's the Difference Between a Trademark, a Copyright and a Patent?

There are three types of intellectual property that you probably know about: Trademarks, copyrights and patents. There's a fourth type, called trade secrets, that very few people know much about. I guess that why they're called trade secrets. We will dissect each of the four types in this column.

A trademark is a word, phrase or symbol that identifies the source of a product and distinguishes it from others. Brand names are trademarked. Trademarked product names or company names are shown with the TM symbol, usually written in a smaller font.

EXAMPLE: *Invention Mysteries* ™

A **registered trademark** is a trademark or service mark which has been registered with the U.S. Patent & Trademark Office (U.S. PTO). The symbol for a registered trademark is ®.

EXAMPLE: Pepsi ®

Trademark rights arise from either using the mark in public OR from filing an application to register it with the PTO. Certain items are not eligible for a trademark, such as letters, numbers, slogans, colors (the pink color of Owens-Corning's insulation).

A **service mark** is the same as a trademark except that it identifies and distinguishes the source of a service rather than a product. Service marks are shown with the symbol SM.

EXAMPLE: GE's *We Bring Good Things To Life SM*

A **copyright** protects an original artistic or literary work; a copyright lasts for the "life of the author plus 70 years." Copyrights protect the following: Literary works, such as books, plays, articles or poems; songs (without a copyright, there would be no royalty payments to the musician); movies, including movie soundtracks; pictures and paintings; architectural works and pantomimes.

A copyright is created when the work is published with the copyright symbol © and the year next to it, and the copyright holder usually places his name on the copyright notice, too. A copyright can also be filed with the Library of Congress to prevent or resolve future disputes over ownership.

EXAMPLE: Warner Brothers © 2001 or Copyright © Warner Brothers 2001.

What is not eligible for copyright protection? Works that consist entirely of common knowledge and contain no original work, such as calendars, rulers, height & weight charts, tape measures.

A **patent** protects an invention for 20 years from the date the patent is applied for with the patent office. There are 3 types of patents: utility, design and plant. A utility patent protects the functionality of an invention, a design patent protects the appearance of an invention and a plant patent, as the name implies, refers to the discovery or creation of a new plant. The patent number, or the term "patent pending," is placed somewhere on the product or the packaging.

The term **"patent pending"** means that a patent has been applied for but has not yet issued. If the patent later issues, the patent holder is protected for twenty years from the date of his application. The unauthorized use of another's patent, trademark or copyright is called "infringement." When this happens, the result is usually litigation through the courts.

Approximately 20 percent of the patents issued each year go to independent inventors while 80 percent go to corporations, yet more than 2/3 of the major new product breakthroughs in the 20th century came from individual inventors rather than corporations. Further, while many people assume that a patent usually makes an inventor wealthy, fewer than 2 percent of all patents actually produce a profit for the inventor.

The fourth and final type of intellectual property is the **trade secret.** Examples of trade secrets include the recipe for Colonel Sanders' chicken and the formula for Coca-Cola ®. Trade secrets are not patented for two reasons: First, patents expire after twenty years. Second, patents become common knowledge once they're issued and, even though they offer legal protection against infringement, patent attorneys and product developers can sometimes "design around the patent." In some cases, a firm will try to reverse engineer a product to find out what ingredients it contains and how it is made.

Now you know the difference between a trademark, a copyright and a patent and what all those symbols stand for.

Could this be the greatest invention of all time? You decide ...

This article tackles that age-old question, "What's the greatest invention of all time?" Since this topic has been debated for many years without any kind of consensus, there may be no one right answer. At the end of the article, you get to tell us what YOU think is the greatest invention of all time.

First, we establish some criteria for selecting the greatest inventions of all time.

We're not nominating any Rube Goldberg inventions for the title of greatest invention. We're also not considering developments such as fire, the wheel, the alphabet or the spoken language because, for the most part, these are considered to be "discoveries" rather than "inventions." Electricity could be classified as a discovery, too, but we include it in this column because of the subsequent electrical inventions that came later.

The three criteria that we'll use to determine the greatest invention of all time are:

1. The number of people who use it or benefit from it.
2. Its impact on society. For example, does it save lives?
3. Its place on the historical timeline: Would this invention be possible without a previous invention?

The top contenders, in no particular order, are:

- Johannes Gutenberg's printing press (invented in the mid-1400's)
- The discovery and use of electricity
- Indoor plumbing (early records place its origin between 2,500 B.C. -- 1,700 B.C.)
- Alexander Graham Bell's telephone (1876)
- Thomas Edison's incandescent light bulb (1879)

- Sir Alexander Fleming's penicillin (1928)
- The mass-produced automobile
- Computers
- The Internet

Using a process of elimination, Criteria # 3 (its place on the historical timeline) eliminates the Internet because it would not exist without the discovery of electricity, the invention of the telephone and computers. Computers cannot be considered the most important invention of all time because they depend on electricity. As a result, electricity is more important than either computers or the Internet.

While the light bulb and the telephone have each been considered by many to be the greatest invention ever, neither one would have been invented without electricity. So these two get voted off the island for the same reason as computers and the Internet.

So what is the single most important invention in history? Obviously, there's no easy answer to this age-old question, but we've narrowed it down a bit by eliminating a few of the contenders. In the next article, we'll discuss the remaining inventions and see which ones get voted off. In the meantime, try to envision what life would be like without the remaining great inventions: the printing press (which brought us type-written books), electricity, indoor plumbing, the automobile and penicillin.

The Debate Continues ... What's the Greatest Invention of All Time?

(This is the second article in a three-part series)

We continue our discussion on what is the greatest invention of all time. Keeping in mind that there may be no one "right" answer, as different people have different ideas for the greatest invention, we base our debate on three criteria:

1. The number of people who use it or benefit from it.
2. Its impact on society. For example, does it save lives?
3. Its place on the historical timeline: Does the invention require the use of a previous invention?

The debate began with these 9 top contenders...

- Johann Gutenberg's printing press in the mid-1400's
- The discovery and use of electricity
- Indoor plumbing
- Alexander Graham Bell's telephone in 1876
- Thomas Edison's incandescent light bulb
- Sir Alexander Fleming's penicillin
- The automobile
- Computers
- The Internet

... and we whittled it down to five: The printing press, electricity, indoor plumbing, the automobile and penicillin. These have all impacted millions of lives in a positive way, so how do we determine which invention has benefited people the most? Which invention is the greatest of all time?

Since none of these inventions required the use of a previous invention, we can eliminate the third criteria (its place on the historical timeline) and focus on the first two criteria: 1) The number of people who use it or benefit from it and 2) Its impact on society.

Sir Alexander Fleming discovered penicillin in 1928 by accident, but it wasn't until two other men, Howard Florey and Earnest Chain, figured out a way to make it fight the spread of bacteria before people could actually benefit from it. In the mid-1940's penicillin became widely available for use as medicine, and now plays a major role in treating illnesses such as pneumonia, rheumatic fever and scarlet fever. In addition, the discovery of penicillin was the foundation for discovering many other antibiotics that are used today.

Penicillin has saved millions of lives, including the lives of thousands of soldiers who landed at Normandy in World War II. Fleming, Florey and Chain shared the Nobel Peace Prize in Physiology & Medicine in 1945. Is penicillin the most important invention in history? If not, then what is?

The Verdict is in: This is the Greatest Invention of All Time

(This is the third article in a three-part series)

One way to measure the value of an invention such as indoor plumbing, which is often taken for granted except when it doesn't work, is to evaluate the quality of life without it. Prior to the widespread use of indoor plumbing, many people died of dysentery, cholera and other sanitation-related diseases. Without indoor plumbing, in addition to using outhouses and not having a healthy way of disposing of waste, we'd have to wash our clothes outside in a tub, bathe in a tub and heat the water that we use for cooking every day. So in addition to being an invention which saves lives, indoor plumbing does away with the inconveniences that would take up valuable time each day.

There are no exact figures on the number of lives that have been lost due to a lack of indoor plumbing, but it's been said to be possibly in the millions worldwide. While approximately one percent of the people in certain southern states in the U.S. are still without indoor plumbing – and millions more in underdeveloped countries, imagine what it would be like living in a big city without it. Okay, don't imagine what it would be like, but you get the picture. In terms of an invention's impact on society and its ability to save lives, I believe indoor plumbing is even more important than penicillin.

Many of today's major inventions would not have been possible if the inventor had not received a good education. Johannes Gutenberg's printing press brought us movable type and type-written books in the mid-1400's, and is considered by many to be the greatest invention ever. Prior to the printing press, only the nobles and the wealthy had access to the kind of education that books afforded. Johannes Gutenberg made education available to the common man when he created his printing press, just as Henry Ford made automobiles available to the common man when he developed the concept of assembly line production.

The mass production of the automobile has changed the world in many ways, but not to the extent that electricity has. While Ben Franklin is widely recognized as the person who discovered electricity in 1752, there were others before him who contributed to it, and many since Franklin have expanded its uses, including one of the greatest inventors of all time, Thomas Edison. We all know that electricity has led to the development of everything from street lamps, indoor lighting, refrigerators and other household appliances, radio and television, the power to run our homes and workplaces, telephones, computers and the Internet, just to name a few. Not much else needs to be said about the importance of this invention.

So in terms of an invention's impact on society AND the number of people who benefit from it, electricity is a greater invention than the automobile. We've already placed indoor plumbing ahead of penicillin, and so the greatest invention of all time is ... we have a tie.

In terms of the number of people who use it or benefit from it, a case can be made for each of the following as the greatest invention of all time: The printing press, indoor plumbing and electricity. Keep in mind that there are millions of people in underdeveloped countries who do not benefit from any of these three inventions.

In terms of its impact on society, all three inventions have had major impacts on society to the point where it's nearly impossible to judge, but indoor plumbing has probably saved the most lives of all three.

The unscientific tie-breaker that we'll use takes into consideration the later inventions that became possible as a result of the invention. That makes electricity the winner, as it has allowed for the development of more great inventions than either the printing press or indoor plumbing, including four of the other all-time great inventions listed earlier: The telephone, the incandescent light bulb, computers and the Internet.

The five greatest inventions of all time, in my humble opinion, are:

1. Electricity
2. Printing press
3. Indoor plumbing
4. Penicillin
5. Automobile

Regardless of what you believe is the greatest invention of all time, there will always be additional inventions created in the future that will have people still debating this topic a hundred years from now!

The Case of the Missing Monkey

Our story begins in 1956 with a lady named Bette Nesmith, who was a typist for a bank. She wasn't a very good typist, though, and found herself erasing her frequent mistakes with a pencil eraser. When her employer switched over to electric typewriters, she could no longer erase her mistakes on the new type of ribbon, so she set out to invent a product that would allow her to paint over her mistakes. Bette created the invention of Liquid Paper, which was originally known as Mistake Out, to help her in her job.

Bette experimented with many different combinations of ingredients until she found the right formula. Later, after she had already begun selling bottles of her Liquid Paper, she was fired from her job. The firing turned out to be a blessing, though, as it enabled her to devote all of her time to her Liquid Paper business. Sales began to skyrocket and, in an attempt to fill the demand while keeping her costs down, Bette had her son, Robert, help her fill the bottles of Liquid Paper.

By 1975, her company, the Liquid Paper Corporation, employed 200 people and produced 25 million bottles of Liquid Paper that were sold in 31 countries. She later sold her company to Gillette for $47 million plus royalties.

I don't mean to change the subject, but back in 1997, the 1960's pop band, *The Monkees*, staged their 30th Anniversary Reunion Tour. Only three of the original four band members participated. Davy Jones, Micky Dolenz and Peter Torkinson, a.k.a. "Tork," were all there, but what happened to the fourth *Monkee*? Why would he skip their reunion tour?

Could it be because the missing *Monkee* was in a much better financial position than the other *Monkees*? If so, how did he become wealthier than the others?

Remember when I mentioned that Bette Nesmith's son, Robert, helped her fill the bottles with Liquid Paper in the 1950's?

Robert is his first name, but he goes by his middle name ... Michael. You probably know him as Michael Nesmith, guitarist for the band *The Monkees* ... Bette's boy.

Congratulations ... you've just solved the "Case of the Missing *Monkee*."

Necessity is NOT always the mother of invention

"It is really through her that I have become competent in the subject (of relativity),"
– Albert Einstein, speaking of Emmy Noether.

Who is Emmy Noether, you ask? She's the woman who devised the mathematical principle called Noether's theorem which helped form the basis of quantum physics. Einstein formulated his general theory of relativity based upon her calculations.

Ladies, you're going to love this article ... and guys, you might learn a thing or two about the fairer sex in this article as well. We'll take a look at the importance of female inventors, including the ways in which they're different from male inventors and how they're similar.

First, consider these lopsided statistics about women inventors:

- In 1890, fewer than one percent of U.S. patents were issued to women.

- There have been over three hundred recipients of the Nobel Prize in the sciences in its 102-year history (one award is presented annually each for chemistry, physics and medicine). Only ten recipients – about three percent – have been women. Marie Curie was the first woman to win the Nobel Prize in physics.

- In 2002, only 15% of the 100,000 utility patents issued to independent inventors that year were issued to women.

This last stat reflects a dramatic improvement over the last hundred years. This is based partially on the fact that during parts of the 1800's, women were not allowed to own property – including patents – once they got married. There were other factors that contributed to this disparity as well. Fewer women than men went to college in the 1800's and early 1900's and women, to this day, face the stereotype that they are not inventors, even though this is constantly being proven to not be true.

There is no evidence to suggest that men are better inventors than women. In fact, just the opposite may be true, as women have a higher percentage of SUCCESSFUL patents than men do. By success, I mean that the patent produced a profit for the inventor. Why is this?

One possible reason, according to Joanne Hayes-Rines, who has been the editor of *Inventors' Digest* magazine (www.InventorsDigest.com) since 1987, is that women tend to be better at marketing their inventions and in working better with others. "Women inventors succeed more at getting their products to market because they're more practical, they define the need better, and they package it better than men. They see the big picture better," says Hayes-Rines.

Whether you agree or disagree with her, there aren't too many people who have more experience working with inventors than Hayes-Rines, as she has made herself into an expert on the subject of inventions. It's hard to tell who in her family knows more about inventions, because her husband, Bob Rines, is a successful patent attorney and inventor himself. He founded the Franklin Pierce Law Center, and is a noted expert on the mystery of the Loch Ness Monster. He's well known for developing the sonar that led to the creation of the technology that was used to discover the Titanic and to hunt for Nessie.

One thing that female inventors have in common with male inventors is that they both create solutions to problems. Having worked with inventors for the past five years myself, I've noticed that people tend to invent products in the industries in which they work – their areas of expertise. Among women, nurses invent products that help them provide better care for their patients, stay-at-home moms invent products that help them raise their kids better and female hair stylists tend to invent products that pertain to hair. Among men, mechanics invent new tools, carpenters invent new construction-related products and coaches invent new products to help their athletes perform better.

Albert Einstein was impressed with the contributions made by women inventors, and I think you should be impressed, too.

In the next column, we showcase three well-known products that were invented by women.

Necessity is not the mother of invention ... these women are

"Very learned women are to be found, in the same manner as female warriors; but they are seldom or never inventors."
– Voltaire

What do windshield wipers, COBOL and Scotchgard have in common?

All were invented by women.

In the previous column, we looked at some of the similarities and differences between male and female inventors. In this article, we take a look at the above three inventions that have had a major impact on society. In addition, here are a few other well-known products created by female inventors that will be featured in future columns:

- Bulletproof vests
- Fire Escapes
- Laser Printers
- Flat-bottom grocery bags
- Certain drugs that fight diseases such as childhood leukemia, herpes, gout and the first drug to fight AIDS drug, AZT

By the way, how many of the inventors of the above products can you name?

Windshield Wipers:

Some inventions are created as the result of a person simply trying to solve a problem. That's what Mary Anderson of Alabama did in 1903 when she invented windshield wipers. On a trip to New York City, while touring the city on a streetcar, she noticed that the motorman had to continually get out to wipe the snow and ice from the windshield.

The man had tried a variety of solutions to this problem but nothing had worked.

After making a quick drawing in her sketchbook, Mary came up with a solution to the problem. Her solution, which would be patented a year later, allowed the motorman to sweep the snow and ice away with a device that was operated from inside the car. This became the forerunner to the modern windshield wiper. Even though wipers had become standard equipment on American cars by 1913, Mary never profited from them.

The *U.S.S. Hopper*:

Grace Murray Hopper developed COBOL (which stands for Common Business Oriented Language) in 1959 while she was in the Navy, and she was also the Navy's first female admiral. COBOL was more like natural English than any previous computer language. It was the first programming language mandated by the Department of Defense for its applications and, in recognition of her contributions, the Navy named one of their destroyers in her honor, the *U.S.S. Hopper*.

COBOL served as a foundation for later computer languages and it's likely that we wouldn't have the World Wide Web today if it weren't for COBOL. Another contribution that Grace Hopper made was the term "computer bug." No, she didn't invent it, but she is the one who coined the term. She did this when a computer processor had stopped working due to a moth that was stuck in it.

Scotchgard:

Some new products are the result of an accident. Penicillin, Post-It Notes®, Silly Putty® and Ivory Soap® are all examples of accidental discoveries. You can put Scotchgard in this category, too.

Patsy Sherman created Scotchgard® in 1952 while working as a chemist for 3M in Minneapolis. Sherman's team had been trying to develop a new kind of rubber for use in aircraft fuel lines when an assistant in her chemistry lab accidentally dropped a beaker full of a liquid rubber mixture onto the floor, splashing onto Sherman's white canvas sneakers.

When they tried to wash it off, the water and solvents beaded up and ran off the sneakers. Sherman and fellow chemist Sam Smith realized that the mixture could be used to protect fabrics from water and other fluids. After three years of work, the mixture was patented and released as Scotchgard Protector ™ in 1956.

In a 1997 speech to students, Sherman explained that being an inventor does not require a lot of money or education, nor is it a matter of age or gender. She once remarked, "How many great discoveries would never have occurred were it not for accidents?"

At this point, what have we learned about the abilities of women inventors?

Is it that women would achieve MORE success than their male counterparts if they had the same opportunities as men? Or LESS success? Or are women and men equal as inventors? The answer to that question is as tough to figure out as removing red wine from a couch that wasn't protected by Patsy Sherman's invention.

Invention Mysteries Pop Quiz:

I'd like to honor the group of students in the marketing class that I teach at Quincy (IL) University in my hometown by turning this article into a pop quiz. You see, I told them recently that they would have a quiz on the "next chapter" the following week.

They failed to read the class syllabus, though, as college students sometimes tend to do. If they had read it, they would have found that we were going straight from Chapter 3 to Chapter 5, skipping Chapter 4 in the process. As a result, they studied the wrong chapter and most of them failed the quiz!

As I tell my students, there are no trick questions on my quizzes. Grading is as follows, and the answers appear at the end of the column. No peeking!

All 15:	Inventive genius
11 – 14:	Wise as an owl
7 – 10:	Average
4 – 6:	Novice
0 – 3:	Go back to the drawing board

1. What protects an invention from infringement for twenty years?
    a. patent
    b. copyright
    c. trade secret
    d. trademark

2. Which "bright" inventor was quoted as saying, *"Genius is one percent inspiration and 99 percent perspiration"*?
    a. Leonardo da Vinci
    b. Thomas Edison
    c. Alexander Graham Bell
    d. Rube Goldberg

3. Which famous invention did Mark Twain turn down as an investment opportunity because he had recently invested in other inventions that had failed?
   a. the telegram
   b. the telestrator
   c. the telephone
   d. the telegraph

4. Possible trick question here ... TRUE or FALSE: The inventor who perfected the Braille alphabet was ... Mr. Braille.

5. Which of the following inventors was born in the late 1300's?
   a. Samuel Morse
   b. Johannes Gutenberg
   c. Joseph Guillotine
   d. Rube Goldberg

6. The only U.S. President to receive a patent was ...
   a. George Washington
   b. Abraham Lincoln
   c. Teddy Roosevelt
   d. Ronald Reagan

7. Where is the U.S. patent office located?
   a. Chicago, IL
   b. Arlington, VA
   c. New York, NY
   d. Kokomo, IN

8. TRUE or FALSE: Around 1899, the former commissioner of the United States patent office was quoted as saying, *"Everything that can be invented – has already been invented."*

9. TRUE or FALSE: Michael Jackson, together with two other inventors, received a patent for an anti-gravity device for his moonwalk shoes.

10. What protects books, plays, articles, songs, etc. from infringement?
    a. patent
    b. copyright
    c. trade secret
    d. trademark

11. TRUE or FALSE: The first U.S. patent to be awarded to a woman didn't happen until the 1900's.

12. Possible trick question # 2: TRUE or FALSE: The inventor of the World Wide Web is an English computer scientist named Timothy Berners-Lee, not Al Gore.

13. TRUE or FALSE: Ben Franklin invented bifocals as well as the first odometer used to measure the routes that mail carriers traveled.

14. TRUE or FALSE: Thomas Edison holds the record for being granted the most U.S. patents for his inventions, with more than 1,000 patents in his name.

15. The famous invention cartoonist who designed elaborate methods to accomplish simple tasks was ...
    a. Rube Goldberg
    b. Rube Goldberg
    c. Rube Goldberg
    d. All of the above

ANSWERS: 1-a; 2-b; 3-c; 4-true; 5-b; 6-b; 7-b; 8-true; 9-true; 10-b; 11-false; 12-true; 13-true; 14-true; 15-Rube Goldberg

What does this man of peace have to do with dynamite?

"If I have a thousand ideas and only one turns out to be good, I am satisfied?"

This is the story of an inventor who held more than 350 patents in his lifetime, yet the invention that he is most remembered for is responsible for the deaths of thousands of innocent people.

Alfred was born in Sweden in 1837, the son of an inventor who built bridges and buildings in Stockholm. When Alfred was nine, he moved to Russia with his family. His father had hired private teachers so that he and his three brothers could receive the finest education possible. As a teenager, Alfred studied in the United States from 1850 – 1852, and he also visited Paris during this time. It was in Paris that he first learned about nitroglycerin.

Alfred wrote poetry and drama and at one point in his life had seriously considered a career in literature. His favorite subject, though, was chemistry. Remember that as you get closer to the end of this story.

While in his early twenties, Alfred set up a lab in Stockholm in 1859 with his father and younger brother to experiment with nitroglycerin. They saw that nitroglycerin had some advantages over gunpowder and could be used for commercial purposes. As they conducted their experiments over the years, there was the occasional lab explosion. Later, in 1864, one of these explosions killed his brother and several other people.

By 1866, Alfred had invented dynamite when he was just twenty-nine years old. He had achieved a far greater level of success than most inventors his age. He built laboratories in more than twenty countries all over the world and eventually held more than 350 patents. His patents included synthetic rubber, leather and artificial silk. The company that he bought in 1893 is today known all over the world as a manufacturer of munitions and firearms.

With all his success, Alfred liked the idea of some day giving away his fortune. Maybe he felt a need to atone for the fact that his most famous invention -- dynamite -- was sometimes responsible for the deaths of innocent people. He established an annual prize to reward those who make the biggest contributions to society each year. What types of contributions did he reward?

They are divided into five classifications:

- physics
- chemistry
- physiology and medicine
- literature
- peace

If this list looks familiar to you, it's because the Nobel Prizes contain the same five classifications of prizes.

Yes, the person who invented dynamite is the same person who is responsible for the Nobel Peace Prize... Alfred Nobel.

Nobel died of a cerebral hemmorrhage in his home in San Remo, Italy, in 1896. The prizes that bear his name were established four years later as he had specified in his will.

Would Mark Twain have preferred to be an inventor rather than a writer?

"Inventors are the creators of the world – after God"
– Mark Twain

While we're all familiar with the writings of Mark Twain, many people don't know that he was highly involved with inventions, both as an inventor himself and as an investor in other people's inventions. Twain profited from some of his own inventions, but he lost a lot of money investing in other people's inventions.

His first invention was for a vest strap that served as a collar and vest, designed to replace suspenders. He filed a patent application for it in September of 1871, but he encountered a problem in getting a patent because an inventor from Baltimore had created a nearly identical product.

"The first thing you want in a new country is a patent office. A country without a patent office and good patent laws ... couldn't travel any way but sideways or backwards"
– Mark Twain

In order to determine who should be awarded the patent, the Commissioner of Patents must institute an "interference," which is a contest to determine who created the invention first. The Commissioner asked each inventor to file a paper listing the essential dates and facts about his invention; the patent would then be awarded to the person who can show that he was the first one to create the invention.

Rather than simply reciting the facts in numbered paragraphs as was the custom back then, Twain sent a handwritten letter in short story form. In his letter, he explained the details of how he created his invention, including the fact that his brother witnessed the exact date of his invention. The Commissioner of Patents probably enjoyed the human element in Twain's letter, as it was a welcome departure from the usual mundane papers submitted to him, but it's not known whether or not this actually helped Twain. The dispute was later resolved when the patent attorneys agreed to settle it based upon the dates in which the applications were filed, rather than the dates in which each man originally created his version of the invention, as U.S. patent law specifies. Twain was granted Patent # 122,992 in December of 1871.

Mark Twain received two other patents during his lifetime. One was for a self-pasting scrapbook in 1873 which he named Mark Twain's Scrapbook, and the other was in 1883 for a game called "Memory Builder." This game made it easier to remember historical dates, but it didn't succeed commercially.

Twain earned a fortune and gained international fame from his writings, yet there was something significant about his scrapbook invention. What was it? Read on; the answer is at the end of the story.

> *"We are called the nation of inventors. And we are. We could still claim that title and wear its loftiest honors if we had stopped with the first thing we invented, which was human liberty"*
> – Mark Twain

While Twain had profited from some of his inventions, there were other inventions that he thought of but did not commercialize. In his notebooks, Twain recorded ideas for microfilm in 1885, and for an invention that would utilize "pictures transferred by light," similar to modern television, in 1888, as well as an idea for the use of fingerprinting, which was the cornerstone of the plot in his novel, *Pudd'nhead Wilson*, published in 1894.

Twain lost more than a half million dollars in his lifetime from the failed inventions in which he invested, including the Paige typesetter. But the invention that cost Twain the most was one in which he did *not* invest in: Alexander Graham Bell's telephone. When Bell personally offered Twain a chance to invest in his telephone, Twain responding by telling him that he wasn't interested because he had been burnt once too often on inventions.

What was the significance of the scrapbook that Twain invented? He earned more money from it that year than he did from his writing. Mark Twain was a moderately successful inventor; without the recognition that he earned from his writings, the world probably would never have known of his inventions. He did, however, achieve more success as an inventor than most inventors do.

It was quite a year for this "mother of invention"

The year was 1965. A half gallon of milk cost just 53 cents, delivered to your front door. Miniskirts were in fashion, Sonny & Cher's *"I Got You, Babe"* was a hit song and *"Lassie"* was one of the most popular shows on TV. The Pillsbury Doughboy made his debut that year, and baseball was played inside the Houston Astrodome for the first time – on natural grass. The Vietnam War continued, as did the war protests, and Martin Luther King, Jr. led marches to protest unfair voter-registration rules of the day. Lyndon Johnson was President.

The year 1965 was also the year that your humble columnist was born. More importantly, though, 1965 was the year that a chemical engineer named Stephanie Kwolek accidentally invented what became known as Kevlar ®.

Working for DuPont in 1965, Ms. Kwolek's boss had asked her to develop some new synthetic polymers. Each day she would mix different combinations of liquid crystals in an attempt to produce new types of fibers, when one day something different happened. She came up with a solution that was cloudy like thick milk, unlike other liquid crystals that are transparent and clear.

When she asked the man in charge of spinning the polymers to run it through the spinneret, which is used to make synthetic fibers, he initially refused because he thought the new solution would plug up the device. After three days of trying to convince him to test it, he finally relented. The brand new "aramid" polymer that she had just invented was a stiff, lightweight material that was five times stronger than steel, and when she baked it, it became even stiffer. The resulting product, Kevlar, is the material used in bulletproof vests which has saved the lives of more than 2,000 police officers since it was introduced for sale six years later.

Stephanie Kwolek's career as a chemical engineer almost came about by accident, too, according to the Lemelson Center at the Smithsonian Institution. "I did not start out to be a chemist. As a child, I thought that I might be a fashion designer," Stephanie recalls.

Stephanie's father had helped her develop an interest in science, but it was her mother who fueled her passion for fashion. As a little girl, Stephanie spent a lot of time sewing and drawing various types of clothing. After college, she went to work for DuPont with plans to stay only until she could raise enough money to go to medical school. It was by pure coincidence that the group she joined at DuPont was the Textile Lab, which was devoted to working with textiles and fibers.

Today, in addition to being used in bulletproof vests, Kevlar is used to make items as diverse as crash helmets, skis and radial tires, and it's also strong enough that it's used in aerospace applications. For her work, Kwolek has been honored by numerous police organizations nationwide, and in 1995 she became only the fourth woman to be elected to the National Inventors Hall of Fame. In 1999 she won the Lemelson-MIT Lifetime Achievement Award. Kwolek isn't a one-hit wonder, though, as her name appears on 16 additional patents issued between 1961 and 1986. Today she consults with DuPont and travels the world speaking to other scientists about her work, and she also mentors young students in science, especially young women.

The invention that nearly ruined its inventor

"I don't think the goal was the magnitude of the money. My role was to defend the patent system"
– Robert Kearns

Occasionally you hear a story in the news about an inventor having his idea stolen by a big company; this is one of those stories. It's about an individual inventor and the battles he fought with the automakers to prevent them from stealing his invention. In 1964, Robert Kearns had invented intermittent windshield wipers while working out of his garage, and his efforts to launch them onto the American market led to court fights that lasted for years.

Most inventors will tell you that it's very hard to introduce a new invention to the automotive industry; Kearns found out the hard way. There are two common ways for an inventor to launch a new invention onto the market: You can license it to an existing company, or manufacture and sell the item yourself.

In an eerie coincidence, as noted in an earlier *Invention Mysteries* story, the first windshield wipers were invented in 1903, exactly one hundred years ago, by Mary Anderson of Alabama during a trip to New York.

For a product like intermittent windshield wipers, there were several reasons why it made more sense to try to license it to an existing company such as Ford, General Motors or Chrysler. First, these companies already have worldwide distribution established; second, they can install the wipers as standard equipment; and, third, it would be nearly impossible for an inventor to achieve critical mass selling the wipers himself. Unfortunately, any of the Big Three automakers could design their own version of intermittent windshield wipers by designing around Kearns' original patent –

and put him out of business.

Accoring to David Lindsey's book, *House of Invention*, when Kearns met with representatives from Ford to demonstrate his wipers, he was told that all he needed to do was prove that they met industry standards. He did this in 1964 but rather than immediately license his wipers, Ford offered Kearns a job instead, which he gladly accepted.

Kearns was laid off just five months later, though, and he soon noticed that his intermittent wipers began appearing on Ford cars, even though he did not license them to Ford. Kearns still held the patent rights. It's one thing to have a company steal your idea, but imagine what it must feel like to have AN ENTIRE INDUSTRY steal your idea! That's what happened to Kearns, as General Motors, Chrysler, Saab, Volvo, Honda and Rolls-Royce all followed Ford's lead in stealing Kearns' intermittent wipers.

Kearns decided to fight them in court, initially serving as his own lawyer even though he had no legal background. Years of legal battles followed, and Kearns eventually won court settlements of $10.2 million from Ford and $11.5 from Chrysler. Today, nearly all new cars sold worldwide have intermittent windshield wipers. So this story turned out just fine for Kearns, right?

Not exactly. The legal battles consumed nearly thirty years of his life. Kearns' daughter Kathy once said, "The lawsuit is all we've ever known." The inventor's wife, after having finally lost her patience, left him. Kearns filed additional lawsuits against nineteen foreign automakers but lost, and his suit against General Motors was thrown out. Altogether, he spent nearly $8 million dollars in legal fees.

We now know that inventor Robert Kearns won the battle, but only he and his family can decide who won the war. His fight was more about inventors' rights than it was about money. It was about principle, as evidenced by the fact that he had turned down an earlier settlement offer of $30 million from Ford. Kearns' intermittent windshield wipers have benefited car owners worldwide, even though the majority of them do not result in a royalty to Kearns.

Kearns wasn't the first inventor who invented a revolutionary new automotive product and didn't see instant rewards. You see, Mary Anderson, who invented the precursor to Kearns' invention, the windshield wiper, in 1903, never profited from her invention even though it had become standard equipment on American cars by 1913. Her story was quite different, though, as it didn't involve lawsuits. Some companies see infringement lawsuits as merely a cost of doing business, but in this case, the automakers nearly ruined an inventor's life in the process.

Take a Flight Back in History to 1903

"Before the Wright Brothers, no one in aviation did anything fundamentally right. Since the Wright Brothers, no one has done anything fundamentally different."
  – Darrel Collins, U.S. Park Service,
   Kitty Hawk National Historical Park

Leonardo da Vinci (1452 – 1519) had envisioned a flying machine nearly five hundred years ago, but it wasn't until the Wright Brothers made a working model of the first airplane in 1903 that human flight was officially invented in the form of their Wright Flyer. The Wright Flyer had a wingspan of 40 feet and weighed a little more than 600 lbs.

This year we celebrate the 100th anniversary of what was then a truly new invention –powered flight. The Wright Brothers had begun glider experiments in 1900 in Kitty Hawk, NC, and in 1902 had conducted more than 1,000 test flights.

Their first powered flight, with Orville at the controls, was on December 17, 1903. It lasted 12 seconds and covered 120 feet. Wilbur piloted the fourth flight later that day, covering 892 feet and staying aloft for 59 seconds. The Wright Brothers survived the day, but their plane didn't, as it was overturned by a gust of wind and destroyed.

On May 22, 1906, the Wright Brothers received a patent for their "Flying machine with a motor." Interestingly, there were a few other important events that happened during this week in aviation history.

- In 1819, the first bicycles -- called swift walkers -- were introduced to the United States in New York City. How is this relevant to flight? As you probably already know, Orville, 36, and Wilbur, 32, owned a bike shop which allowed them to pay the bills while they made aviation history.

- On May 21, 1927, Charles Lindbergh, who was only 25 years old at the time, made aviation history when he flew the first solo flight across the Atlantic Ocean, from Long Island, N.Y., to France in his *Spirit of St. Louis*. It took him 33 hours.

- On May 20, 1932, Amelia Earhart flew from Newfoundland to Ireland to become the first woman to fly solo across the Atlantic Ocean.

- This same week in 1939 saw regular transatlantic air service begin as the *Yankee Clipper* took off from Port Washington, N.Y., to Europe.

Flight has changed the world for the better by making visits to foreign countries possible for millions of people, but it also changed the world for worse by allowing for quicker destruction of human life during wars and terrorist attacks.

There are two rather interesting facts about flight that remain true one hundred years after Orville and Wilbur's first powered flight – one good and one bad. The first is that flying, despite its risks, is statistically much safer than driving, and the other is that most of the major airlines are losing money. American Airlines is bordering on bankruptcy, and TWA filed for bankruptcy last year. This shouldn't come as much of a surprise to those who are familiar with the plight of the Wright Brothers, as they didn't get wealthy from being aviation pioneers. The Wright Brothers faced expensive lawsuits from copycat inventors who tried to infringe on their patent. Even though they eventually won, their legal battles were expensive and time-consuming.

According to the web site www.wright-brothers.org, most of the money to be made was in exhibition flying, where the audiences wanted to see death-defying feats. The Wright Brothers' teams of pilots began to die in accidents and the stress began to affect the Wright Brothers. This, combined with their legal troubles, distracted them from what they were best at –

invention and innovation. By 1911, Wright aircraft were no longer the best flying machines and, in 1912, Wilbur contracted typhoid and died. Orville sold the Wright Company in 1916 and went back to inventing.

For more information, and to see a timeline of the Wright Brothers' history, visit www.wright-brothers.org.

Accidental Inventions

Name the greatest of all inventions.

"Accident." – Mark Twain

Some of the more interesting stories about new inventions are the ones that are created by accident. An "accidental invention" can be born when one inventor develops a product and another person finds a use for it; it can be developed by one of the largest companies in the world and marketed by an outside entrepreneur after the company fails to find a use for it; or it can simply have an employee change the composition of it by accident and in the process give it its most important attribute.

Accidental inventions are all around us, and today we examine three popular products with interesting and different backgrounds.

First up are the Post-It Notes ® from 3M. Art Fry was a researcher at 3M in the 1970's who specialized in developing new products when one of his colleagues had developed a certain adhesive for use in 3M's glues, but he just couldn't get it to stick. Four years later, when Fry was trying to come up with a way to bookmark certain pages in his song hymnal at church, he had a "Eureka" moment and thought that using the adhesive that his colleague had earlier developed just might work. It turned out that it was strong enough to hold the bookmarks in place, yet weak enough that it would allow them to be removed when necessary. The rest, as we all know, is history. This was in 1980 and Post-It Notes was chosen as 3M's Outstanding New Product a year later. Millions of Post-It Notes have been sold each year since.

Back in 1944, during World War II, the United States government asked several large companies to try to make a synthetic rubber for airplane tires, soldiers' boots, etc.

James Wright, an engineer at General Electric, developed a new type of rubber substance that bounced. After the war GE tried to find a use for this gooey material but, unfortunately, they couldn't think of any. Four years later, an entrepreneurial shop owner named Peter Hodgson came up with an idea for it, bought the rights to it, packaged it in an egg-shaped container, gave it a fun name and began selling it by the truckload.

What was this accidental invention?

It was Silly Putty, and it has been a big hit in the fifty years since its debut.

Sometimes it's better to be lucky than good

More than 100 years ago, another invention with an interesting story behind it popped up in Cincinnati, Ohio. While Ivory Soap wasn't an accidental invention, the characteristic for which it's known was an accidental discovery.

An employee who was in charge of the soap-making machine forgot to shut it off one day before he went to lunch. The additional mixing time caused the soap to become puffed up with air. Since it didn't affect the soap, they decided to ship it to their customers. Later, much to their surprise, they began to receive requests from customers for more of "the floating soap." The additional air made the soap lighter than air, causing it to float in water.

These are just a few of the many accidental inventions or discoveries that have become popular; others include aspirin, X-rays, frisbees, velcro, penicillin, Coca-Cola and the Slinky, according to the book, *Mistakes That Worked*. One thing that most accidental inventions have in common is that people did not realize they wanted these inventions until they actually saw them being used. They weren't created to solve a specific need, but in the process they filled a need that people didn't even know they had.

So the next time you see a product and ask yourself what the inventor must have been thinking, keep in mind that it might, with a little tweaking, someday become a great invention.

Necessity is the *mother* of invention, so celebrate Father's Day with these *"fathers* of invention"

"One way to predict the future is to invent it."
– Alan Kay

In the next article we begin a three-part series in which we chronicle three of the most successful inventors in United States history. These three inventors are:

• THOMAS ALVA EDISON (1847 – 1931; born in Milan, Ohio) The greatest inventor of all – 1,093 U.S. patents

• JEROME LEMELSON (1923 – 1996; born in Staten Island, New York) The most prolific inventor of our time, with more than 550 U.S. patents

• STANLEY MASON (born in 1921; born in Trenton, New Jersey; still living) Inventor of ordinary, everyday products, with more than 55 U.S. patents, he sold his first invention at the age of seven.

The above three inventors deserve to be called "Fathers of Invention" for the huge impacts their inventions have had on our lives. Between them, they had a combined 1,700 patents, and counting. We'll profile them over the next three articles.

First though, we begin with the story of the person who made it all possible.

There was an inventor born 260 years ago in Virginia who could be considered the original father of invention. Without him, many of our nation's greatest inventions would never have made their way into the public domain.

President Thomas Jefferson invented the most new products of any President in history, yet he chose not to file for patents on any of his inventions. Jefferson was not the one U.S. President to hold a patent -- that would be Abraham Lincoln -- or the one U.S. President to hold a trademark -- that would be George Washington, but what he did do was to help establish our patent system. It was his impact on our patent system that makes him stand out.

Prior to becoming our nation's first head of the patent department in 1790, Jefferson opposed the concept of granting patents because he considered them to be an unfair monopoly.

What's so important about having a good patent system?

Without it, inventors (and companies) would not have as much incentive to create new products, because there would be nothing to prevent other people (or companies) from stealing or copying them. As a result, many of our nation's greatest inventions would never be created. No one would want to invest his own time and money in developing and marketing a new idea if he knew that his idea could be stolen or easily copied by someone else. There must be some sort of protection provided to the inventor in order for him to benefit from his invention, or else there would be less incentive to invent.

Jefferson later changed his mind about the importance of having patents once he saw a number of new products introduced to the general public as a result of the protection that a patent provides. He realized that patent laws INCREASED rather than DECREASED the number of innovations.

Serving as head of the patent office and as Secretary of State at the same time, Jefferson rejected the majority of patent applications that he reviewed. He also established that each new invention must meet three criteria in order to be granted a patent: It must be "new, not obvious, and useful." Without such rules, anybody and everybody who could afford the patent fee would be applying for patents, thus clogging up the system and making it harder for worthwhile inventions to be introduced into society. Jefferson even tested many of the inventions that were submitted to him, and his three criteria for granting a patent have been used in issuing our nation's six million patents.

What are some of the things that Thomas Jefferson invented? The list includes a Moldboard plow, Wheel cipher, Spherical sundial, Portable copying press, Automatic double doors, Bookstand, Swivel chair, Dumbwaiter and a Macaroni machine. In addition, he introduced four food items to the U.S.: french fries, ice cream, waffles, and macaroni.

Jefferson died on July 4th, 1826 – the same day that John Adams died and exactly 50 years after he signed the *Declaration of Independence*.

Invention Mysteries Crossword Puzzle

Across

3. Internet-related invention of Tim Berners-Lee
7. invented the printing press in the 1400's
9. only United States president to receive a patent
10. sport invented by either Abner Doubleday or Alexander Cartwright
12. Cyrus McCormick's invention
15. protects books, plays, articles, songs, etc. from infringement
16. Mary Kies was the first _____ American to receive a patent, in 1809
17. U.S. president who helped establish the patent office
20. famous for theory of relativity
21. inventor of modern air conditioning
23. invented the alternating-current electric motor
24. inventor of the cotton gin
25. word that means a patent has been applied for but not yet issued
29. invented a sewing machine in 1851 that still carries his name
30. frozen foods pioneer
31. invented 8-track tape player and founded Lear jet
32. found more than 300 uses for peanuts and later became head of Tuskegee Institute
33. Philo T. Farnsworth's invention
34. sport that Dr. James Naismith invented
35. 1700's U.S. president who received a trademark

Down

1. invented the telegraph and has a famous code named after him
2. responsible for the guillotine
4. Ruth Handler created this popular doll in 1959
5. protects an invention for 20 years
6. invented the first internal-combustion engine automobile in 1885
8. penicillin
11. Thomas Jennings was the first _____ American to receive a patent, 1821
13. device that Wilson Greatbatch created for heart patients
14. code name for Dean Kamen's "segway" human transporter
18. Most prolific inventor in U.S. history
19. New Jersey home of Thomas Edison's lab
22. winter snow-riding device that Jake Burton invented
26. tire company founder who invented rubber vulcanization in 1839
27. invention that Al Gore did NOT invent
28. Ben Franklin invented these to help people see better

Invention Mysteries Crossword Puzzle

Created by Paul Niemann with EclipseCrossword — www.eclipsecrossword.com

Invention Mysteries Crossword Puzzle

Created by Paul Niemann with EclipseCrossword — www.eclipsecrossword.com

Across and down answers visible in grid:

- GUTENBERG
- WEB
- BASEBALL
- LINCOLN
- REAPER
- COPYRIGHT
- FEMALE
- JEFFERSON
- EINSTEIN
- CARRIER
- TESLA
- WHITNEY
- PENDING
- SINGER
- BIRDSEYE
- LEAR
- CARVER
- TELEVISION
- BASKETBALL
- WASHINGTON

Down words: MORSS, PATENT, PAGE, BIENVENIDO, BEIZ, GRIN, GRINGE, FRICICA, LIFEMING, REAKE, CARES, MANNT, SODA, SNOWBOARD, INTREND, BENFOCAL, GOODYEAR, EDRARD, CANAL

Meet Thomas Edison, the greatest inventor of all time, with 1,093 U.S. patents

(This is the first in a 3-part series on the "Fathers of Invention")

"To invent, you need a good imagination and a pile of junk."
– Thomas Alva Edison

The greatest American inventor of all time was born in 1847 in Milan, Ohio. During his lifetime, Thomas Edison was issued more than 1,000 patents. Some of his inventions – such as the incandescent light bulb and the phonograph – even led to the creation of brand new industries.

"Genius is 99 % perspiration and 1 % inspiration."

Of all of Edison's important inventions, he is best known for the incandescent light bulb, which he created in 1878. Edison wasn't the only inventor to invent a light bulb, though; a British inventor named Joseph Swan developed a different version a year earlier. Trying to find the right filament to make the bulb work was the biggest obstacle Edison faced. His idea for the light bulb was the 1 % inspiration part of the equation, while the thousands of experiments and the marketing of the light bulb made up the 99 % perspiration. The main difference between Edison and Swan was that Edison was able to create an entire industry around his light bulb, which replaced candles and gas lamps as the primary sources of light.

There are several interesting facts about Thomas Edison that many people do not know. For example:

- Edison went to school only until the third grade; then his mother taught him at home. One of Edison's teachers showed how badly she had misunderstood the 6-year-old Edison when she sent a note home with him stating that, "He is too stupid to learn."

- Edison suffered from a hearing loss. This may have helped his career, though, because it allowed him to concentrate better and avoid many of the distractions in his lab.

- Edison once worked sixty hours straight, stopping only for 15-minute catnaps and snacks.

In 1869, Edison was approached by a wealthy businessman about selling one of his products, an improved stock tickertape machine. Rather than stating an asking price, Edison asked the man to make him an offer. The man offered him $40,000 for it (which was equivalent to about $700,000 in today's dollars). This turned out to be a lot more than Edison thought it was worth, and the money helped finance future inventions.

"If we could do all the things we are capable of doing, we would literally astound ourselves."

The incandescent light bulb was Edison's greatest invention. In addition to the light bulb, his 1,093 patents included:

- Vote Recorder (1868)
- Printing Telegraph (1869)
- Stock Ticker (1869)
- Automatic Telegraph (1872)
- Electric Pen (1876)
- Carbon Telephone Transmitter (1877)
- Phonograph (1877)
- Dynamo (1879)
- Incandescent Electric Lamp (1879)
- Electric Motor (1881)
- Talking Doll (1886)
- Projecting Kinetoscope film projector (1897)
- Storage Battery (1900)

Source: Smithsonian Institution

"I have not failed. I have merely found 10,000 ways that won't work."

Even the greatest inventor of all time had a few failures. According to the Inventors.about.com web site, Edison's failures included motion pictures with sound, his inability to create a practical way to mine iron ore, and an electric vote recorder. Even though the electric vote recorder worked, it was a commercial failure and led Edison to remark, "I only want to invent things that will sell."

The electric industry that Edison formed led to the creation of what is known today as General Electric, and his Menlo Park invention lab became the model for which the labs of many innovative companies were patterned after. While Edison's greatest invention was the incandescent light bulb, his greatest contribution is probably the fact that nearly every civilized society on earth has benefited positively from one or more of his inventions.

Meet JEROME LEMELSON (1923 – 1996) – The most prolific inventor of the modern era

(This is the second in a 3-part series on the "Fathers of Invention")

"Don't cut back on basic research. It's necessary to the future of this country."
– Jerome Lemelson, holder of 583 patents

Born on Staten Island in 1923, Jerome Lemelson is one of those inventors who is unknown to 90 percent of the population, yet 90 percent of us use one or more of his products regularly. He received 583 patents and was known as America's most prolific living inventor until his death in 1997.

How long does it take a person to receive 583 patents? If you average one patent per month for 48 years, then you'll come close to Jerome's total. Many of his patents became commercial successes, which is remarkable considering that fewer than 2 percent of all patents ever produce a profit for the inventor!

In all of history, there are only two inventors who have held more patents than Jerome Lemelson: Thomas Edison and Edwin Land, of Polaroid fame. One major difference between them, though, is that Edison and Land both had huge staffs helping them invent. Lemelson had a staff of just one – himself.

According to the Lemelson Foundation web site, these are some of the more well-known products which Jerome either invented or made improvements to: Bank ATM's, the tape mechanisms found in Sony ® Walkmans ®, cordless phones, cassette players and camcorders, fax machines and personal computers, and crying baby dolls. One of his inventions – a universal robot that could measure, weld, rivet, transport and even inspect for quality control -- used a new technology called machine vision.

The technology in his machine vision is what allows bar code readers to scan items at the checkout lines in grocery stores, and it is this technology for which he is most well-known.

I interviewed a friend of the late Jerome Lemelson, Don Costar of Reno, Nevada. Don recalled having lunch with Jerome at a Chinese restaurant one day. He noticed that Jerome's mind began to wander from the conversation they were having. Knowing that Jerome often had two things going on in his mind at the same time, Don asked him what new invention he was thinking about as they were talking.

It turns out that he was trying to figure out a way to improve upon the design of a product which has existed in its current form for hundreds of years: chop sticks!

Lemelson was a big supporter of the rights of the independent inventor against the big companies, having been an independent inventor himself. He took the big companies to court but he lost more often than he won. There were twenty court battles altogether, and these court battles often took him away from what he did best: Invent.

There's an ongoing debate over whether Jerome spent too much time litigating over his patents. Some say that he would invent minor improvements to patents that already existed before companies made those improvements, and then challenge that they infringed on his patents. In a Pete Rose kind of way, this is probably the only reason why Jerome Lemelson is not in the Inventor's Hall of Fame, which is located in Akron, Ohio.

To date, the Lemelson Foundation has pledged $70 million towards its domestic and foreign innovation programs. Its $10 million donation to the Smithsonian Institution was the largest donation the Smithsonian has ever received. The Lemelson Foundation gives out the $500,000 Lemelson – M.I.T. prize each year to the inventor whose works make the greatest contribution to society. The award is known as the "Oscar for Inventors."

When Jerome was diagnosed with cancer in 1996, he did what you would expect him to do; he invented improvements to medical devices and cancer treatments. He submitted nearly forty patent applications during the last year of his life, according to the Lemelson Foundation web site. Some of these patents issued after he died.

Meet Inventor STANLEY MASON, who sold his first invention at the age of seven

(This is the third in a 3-part series on the "Fathers of Invention")

STANLEY MASON was born in 1921 in Trenton, New Jersey. He turned out his first successful invention when he was just seven years old – a clothespin fishing lure that he sold to his friends. Now 82 years young and living in Connecticut, Stanley Mason has created more than 100 inventions in his lifetime and received 60 U.S. patents over the last 50 years.

Stanley Mason had something in common with Thomas Edison and Jerome Lemelson, who we profiled over the last two articles: They each realized early in their inventing careers that there was no use inventing anything which could not be sold at a profit. This is important when you consider that 98 percent of all patents fail to produce a profit.

Yet Stanley is different from Edison in the fact that he considers himself to be "an inventor of ordinary, everyday products." Like Edison but unlike Lemelson, he started out working alone and later established his own team of inventors. Like many great inventors, Stanley improves existing products and creates entirely new ones. Major products that fit into his "new & improved" category include:

- the squeezable catsup bottle
- the underwire bra
- stringless Band-Aid ® packaging
- dental floss dispensers
- "instant" splints and casts for broken limbs

while his most successful "brand new" products include:

- the first granola bars
- fingerprint printing systems

- heated pizza boxes
- heatproof plastic microwave cookware
- masonware
- many toys and games
- the world's first form-fitted disposable diaper with sticky tabs rather than pins.

This last one, the form-fitted disposable diaper with no pins, is the product for which Mason is most well-known. He didn't invent the first disposable diaper, though. That was accomplished by Marion Donovan in 1946, and she later sold her diaper rights to the founder of Pampers nine years later for $1 million.

In case you're wondering, it was NOT Stanley Mason who invented the Mason jar. That honor goes to John Mason of New York City, who patented it in 1858. Like many new products, the Mason jar is named after its inventor. In case you're still wondering about the Mason jar, both the flat metal disk and the lid of the Mason jar were invented by two separate inventors in later years, neither of whom was named Mason.

Now, back to our story.

Mason reportedly starts every morning with an invention and believes in patenting each new idea before telling anybody about it. Some inventors, though, will tell you that it's better to do some market research before patenting your product to determine if there's a market for it.

The company that Stanley started, Simco, located adjacent to his Connecticut farmhouse, has been inventing new products for Fortune 500 companies since its inception in 1973, specializing in food-packaging, cosmetics, and medical devices.

Stanley's wife, Charlotte, is an inventor in her own right. She has received patents and has taught African women in the Congo how to become entrepreneurial.

Stanley and Charlotte have traveled and worked in
80 countries around the world.

Stanley Mason currently teaches entrepreneurship in an MBA
program at Sacred Heart University in Fairfield, CT. He also
lectures at high schools on entrepreneurship and has written
two books on inventing. *"Going Solo"* and *"Inventing Small
Products for Big Profits, Quickly"* are available nationwide at
major bookstores.

As Annika Sorenstam competes on the PGA Tour, find out how golf was invented nearly 500 years ago

"It is no coincidence that the people who invented golf also invented Scotch."
– Bruce Manclark, 1999

The sports media was abuzz over Annika Sorenstam making her debut on the PGA Tour recently. She's the first woman to play on the Tour since Babe Didrikson Zaharias first did it in 1945 – 58 years ago. Watching Sorenstam play, I remembered reading somewhere that the word "golf" is an acronym for "Gentlemen Only – Ladies Forbidden."

Seeing Sorenstam play against the men after some of them complained that she should not be allowed to play, combined with last month's story of Augusta National Golf Club's refusal to accept women as members, would make the g.o.l.f. acronym very interesting – if it's true.

Not knowing whether it was fact or folklore – after all, I had read it on the Internet -- my curiosity got the best of me and I decided to investigate. In the process, I found that there are several competing versions as to when and where golf was invented.

In baseball, there are those who say that the game was invented by Abner Doubleday, whose descendents now run the New York Mets. Others claim that Alexander Cartwright invented it. Either way, there are only two competing versions.

Like baseball, the origin of golf has never been clearly established. Unlike baseball, though, there are four or five competing versions as to when and where the game originated.

The earliest version of golf came from the Romans during Julius Caesar's reign, in which the game was played with a cowhide-type of ball stuffed with feathers and struck with club-shaped branches. There are stories of the Dutch playing on frozen canals around 1425. Variations of golf were also played in France and Belgium. The main flaw with the Dutch and French versions lie in the fact that they lacked at least one essential element of the game – the hole.

Golf as we know it today actually originated in Scotland around 1450. Its exact origins are unknown, but it is believed that golf originated with men AND women along the Scottish coast hitting a pebble with a stick, although the game may have first been played in the Scottish moors by shepherds.

In 1457, King James II temporarily banned golf in Scotland because it interfered with the practice of archery, which was vital to the country's national defense. The residents, though, ignored the ban and began playing on seaside courses called "links," a term still used today to refer to golf courses. King James' son, James III, and his grandson, James IV, also tried to ban golf in Scotland but, like a drunk trying to enforce prohibition, James IV also took up the game.

King James VI of Scotland, who later became known as King James I of England, brought the sport with him from Scotland around 1603. King James' mother, Mary Queen of Scots, also took up the game of golf.

St. Andrews golf course in Scotland is the world's oldest course. A number of 6-, 8-, 9-, and 12-hole courses were opened in the United States around 1890, and the first 18-hole course, the Chicago Golf Club, was founded in 1893.

So where does the word "golf" come from and what does it mean?

It turns out that the word golf is not an acronym at all; it is derived from the Scottish word "gowf," meaning "to strike."

While the "Gentlemen Only -- Ladies Forbidden" philosophy

still forbids women from becoming members at Augusta, this is not the case with the PGA Tour. PGA stands for Professional Golf Association. Nowhere does it state that it is the "men's PGA." The LPGA Tour, on the other hand, states clearly in its title that it's for ladies and, as a result, allows only female players to compete. But I doubt that the women's game will ever be renamed as l.o.m.f., which would stand for Ladies Only – Men Forbidden.

Test your knowledge of inventors and their inventions with this pop quiz

See how many of these inventors you can identify with their inventions. The year of the invention is given, and the answers to the first twenty questions appeared in previous *Invention Mysteries* articles. There are no trick questions, and the answers appear at the end of the column.
Grading is as follows:

27 – 30:	Inventive genius
21 – 26:	Wise as an owl
15 – 20:	Average
7 – 14:	Novice
0 – 6:	Go back to the drawing board

1. Invented the printing press in 1450
2. Invented the bifocal lens in 1780
3. Invented dynamite in 1866; has series of famous awards named after him
4. Invented the talking phonograph in 1877
5. Invented windshield wipers in 1903
6. The person who submitted his own telephone patent two hours after Bell in 1876
7. The only U.S. president to receive a trademark
8. The only U.S. president to receive a patent
9. Person who said, *"Oh, I see they've got the Internet on computers now."*
10. President who said of the telephone in 1876, *"That's an amazing invention, but who would ever want to use one of them?"*
11. Writer who is credited with saying, *"Necessity is the mother of invention"* (what he really said was, *"Invention breeds invention"*).
12. Entertainer whose name is on a patent for anti-gravity shoes (it's not Homer Simpson)
13. Group of people whose code helped us win World War II

14. Inventor of Liquid Paper whose son was a member of the 1960's band, *"The Monkees"*
15. City where the U.S. patent office is located
16. The famous inventor who designed elaborate methods to accomplish simple tasks
17. Writer who earned more money from his self-pasting scrapbook invention in 1873 than he did from his writing that year
18. Country where golf was invented around 1450
19. Post-It Notes ®, Silly Putty ® and Ivory Soap ® all fall into this category
20. New Jersey city where Thomas Edison built his lab
21. Invented the seed drill in 1701
22. Invented the lightning rod in 1752
23. Invented the cotton gin in 1793
24. Invented the reaper in 1831
25. Invented vulcanized rubber in 1839
26. Invented the two-cycle automobile engine in 1879
27. Invented the radio in 1895
28. Invented air conditioning in 1911
29. Invented quick-frozen food in 1924
30. Invented the artificial heart in 1982

**Choose the correct answers from the following.
Each answer is used only once.**

Cyrus McCormick
Mary Anderson
Ralph Waldo Emerson
Johann Gutenberg
Arlington, VA
Navaho Indians
Abraham Lincoln
Benjamin Franklin
Rube Goldberg
Thomas Jefferson
Alfred Nobel
Michael Jackson
Thomas Alva Edison
Menlo Park
Elisha Gray

Jethro Tull
Rutherford B. Hayes
Homer Simpson
Mark Twain
Accidental inventions
Willis Carrier
Scotland
Karl Benz
Bette Nesmith
Charles Goodyear
Benjamin Franklin
Clarence Birdseye
Eli Whitney
Guglielmo Marconi
Robert Jarvik

ANSWERS: 1-Gutenberg, 2-Franklin, 3-Nobel, 4-Edison, 5-Anderson, 6-Gray, 7-Jefferson, 8-Lincoln, 9-Simpson, 10-Hayes, 11-Emerson, 12-Jackson, 13-Navaho Indians, 14-Nesmith, 15-Arlinton, 16-Goldberg, 17-Twain, 18-Scotland, 19-Accidental inventions, 20-Menlo Park, 21-Tull, 22-Franklin, 23-Whitney, 24-McCormich, 25-Goodyear, 26-Benz, 27-Marconi, 28-Carrier, 29-Birdseye, 30-Jarvik.

Take a ride back in time to the 1800's to see how bicycles were invented and re-invented

The year 2003 marks the 100th anniversary of several historic events: Ford Motor Company was founded in 1903, the Wright Brothers' successfully flew for the first time, the first World Series was played between Pittsburgh and Boston, and bicyclists competed in the first Tour de France.

Today we look at the development of the invention that made the Tour de France possible, beginning with the first bicycle invented 86 years earlier.

In this article, you'll see how bicycles played a role in the development of the airplane, motorcycles and automobiles. But first, we take a ride back in history to see who invented the earliest versions of the bicycle and how they've evolved over time.

The Walking Machine – just like the Flintstones did it:

In 1817, Baron Karl von Drais of Germany invented the first version of the bicycle, called the Draisienne. It came complete with a steering bar, but it had no pedals or brakes and was made entirely of wood. Riding it required you to push your feet along the ground one at a time to propel yourself forward.

Kirkpatrick MacMillan, a blacksmith from Scotland, invented the first bicycle with foot pedals in the 1830's to 1840's, but he never patented it and it didn't catch on.

The Velocipede – 100% All-Natural Ingredients:

Making its debut in 1865, the velocipede had pedals applied directly to the front wheel. Like its predecessor, it was made of wood and gave a very rough ride.

**The High Wheel Bicycle – The First One to be
Called a Bicycle:**

In 1870, the first all-metal frame appeared. With rubber tires
and front-wheel spokes, it gave a much smoother ride. This is
the version with the huge front wheel; it was believed that the
bigger wheel would allow you to go faster, and it actually did
allow you to go farther with each rotation of the tires.

The high wheel bicycle was the first one to be called a bicycle,
and they cost an average worker six months' worth of pay. In
1864, the roller drive chain was invented, and is still used on
bicycles today. Ball bearings were first used on bicycles in
1877. Other innovations included the use of a chain with
sprockets and air-filled tires in the 1880's. The pneumatic tire
was invented by an Irish veterinarian named John Dunlop
(as in Dunlop tires) in 1888. The high wheel bicycle was
replaced with the "safety bicycle" in the 1880's, which involved
the use of a chain with sprockets and had two wheels of the
same size.

The High Wheel Tricycle:

The adult tricycle contained two large rear wheels and one
normal-sized front wheel, and was popular with women and
with men who had to wear formal clothing to work.

Some of the mechanical innovations used in cars today were
originally invented for tricycles, such as rack and pinion
steering, differentials and band brakes. Gottlieb Daimler, of
Daimler-Benz fame, mounted his gas engine on a bicycle to
create the world's first motorcycle. The Duryea brothers,
Charles and Frank, were among the first to build a successful
automobile in 1896 and, like the Wright Brothers, they were
bicycle mechanics.

Recent Models:

Three-speed bicycles were popular from the 1950's through the

1970's, until the 10-speed version began to replace them. Today, the latest models are mountain bikes and 24-speed bicycles, and the high-tech bicycles that race in the Tour de France have aerodynamic frames and ultra lightweight carbon fiber wheels.

How did bicycles play a role in the Wright Brothers inventing the first airplane?

The bicycle shop that the Wright Brothers ran before they began flying produced enough income to afford the brothers the opportunity to build and test their airplanes. They learned many of the basics of flight from their experiences in working on bicycles, such as how to transmit power with a chain and sprockets and how to steer. They also used a bicycle when testing their airplanes' wing designs.

So there you have it – a brief story of the long history of bicycles, from the earliest version made of wood that had neither pedals nor brakes, all the way up to the current 24-speed version and the high-tech bicycle that American Lance Armstrong is using to drive the French crazy with his attempt at a fifth straight Tour de France victory.

It's time to de-bunk some invention myths

You may have noticed some of the invention-related myths circulating on the Internet lately. Today we set the record straight on three of the most common ones.

MYTH # 1:
His name was Fleming, and he was a poor Scottish farmer. One day, while trying to eke out a living for his family, he heard a cry for help coming from a nearby bog. He dropped his tools and ran to the bog. There, mired to his waist in black muck, was a terrified boy, screaming and struggling to free himself. Farmer Fleming saved the lad from what could have been a slow and terrifying death.

The next day, a fancy carriage pulled up to the Scotsman's sparse surroundings. An elegantly dressed nobleman stepped out and introduced himself as the father of the boy Farmer Fleming had saved.

"I want to repay you," said the nobleman. "You saved my son's life." "No, I can't accept payment for what I did," the Scottish farmer replied, waving off the offer. At that moment, the farmer's own son came to the door of the family hovel. "Is that your son?" the nobleman asked. "Yes," the farmer replied proudly. "I'll make you a deal. Let me take him and give him a good education. If the lad is anything like his father, he'll grow to a man you can be proud of."

And that he did. In time, Farmer Fleming's son graduated from St. Mary's Hospital Medical School in London, and went on to become known throughout the world as the noted Sir Alexander Fleming, the discoverer of Penicillin.

Years afterward, the nobleman's son was stricken with pneumonia. What saved him? Penicillin.

The name of the nobleman? Lord Randolph Churchill. His son's name? Sir Winston Churchill.

REALITY:
Interesting story that ties together two Nobel prize winners. Too bad it isn't true. Churchill did contract pneumonia in 1943 and 1944, but he wasn't treated with penicillin.

MYTH # 2:
The standard distance between railroad rails in the U.S. is 4 feet, 8.5 inches. That gauge is used because that's the way they built them in England, and English expatriates built the U.S. Railroads.

The English built them like that because the first rail lines were built by the same people who built the pre-railroad tramways, and that's the gauge they used. They used that gauge because the people who built the tramways used the same jigs and tools that they used for building wagons, which used that wheel spacing.

The wagons had that particular odd wheel spacing because if they tried to use any other spacing, the wagon wheels would break on some of the old, long distance roads in England, since that's the spacing of the wheel ruts.

So who built those old rutted roads?

Imperial Rome built the first long distance roads in Europe (and England) for their legions. The roads have been used ever since.

Then what caused the ruts in the roads?

Roman war chariots formed the initial ruts, which everyone else had to match for fear of destroying their wagon wheels. Since the chariots were made for Imperial Rome, they were all alike in the matter of wheel spacing.

The United States' standard railroad gauge of 4 feet, 8.5 inches is derived from the original specifications for an Imperial Roman war chariot, which was made just wide enough to accommodate … the rear ends of two war horses!

REALITY:
This story is also false, according to the web site TruthOrFiction.com, which investigates myths and urban legends. This story started sometime after World War II, but it's not known where it originated.

MYTH # 3:

The design of the white star on Montblanc pens, which
represents the white snow-capped mountain Mont Blanc, was
designed for Adolph Hitler by a Jew, who actually designed it
to resemble the Jewish symbol, the Star of David. As a result,
Hitler carried the Star of David in his pocket without even
knowing it.

REALITY:

The design does resemble the Star of David, but it is probably
just a coincidence. The Montblanc company was founded in
Germany in the early 1900's under a different name, then
became Montblanc in 1910, according to TruthOrFiction.com.
They quote the book, *"The Montblanc Diary and Collector's
Guide,"* which says that the first pens with the white star on
the cap were produced in 1914, which is long before Hitler rose
to power.

What do these three myths have in common?

They're about as real as cow tipping and snipe hunting!

De-bunking more invention myths

In the previous column, we exposed several myths related to inventions; in this article we take on some of the myths that are related to the *process* of inventing.

MYTH # 1:
Inventors are eccentric, frazzle-haired old people who stay up all hours of the night inventing in their garages.

REALITY:

While many people think that the typical inventor looks like Albert Einstein or the gray-haired Doc Brown character played by Christopher Lloyd in the *"Back to the Future"* movies, the truth is that inventors look no different than the rest of us. In fact, there is no such thing as a "typical" inventor. There are several reasons why inventors work out of their garages: Some are mechanically minded, the garage provides them with a better work environment than the living room, and it's cheaper than renting additional workspace.

MYTH # 2:
Men are better inventors than women.

REALITY:
Simply put, there are more male inventors than female inventors. There have been fewer opportunities for women; in fact, there was even a time during the 1700's and early 1800's when the law required that all patents be listed in a man's name, regardless of who invented the item.

Today, women account for fewer than twenty percent of the patents issued in the U.S., but they tend to have more success, on average, than their male counterparts. One possible explanation for this comes from the editor of *Inventors' Digest* magazine, who says, "Women are more organized and tend to work together better. Plus, they tend to be better at the marketing of their inventions." The editor's opinions are based on fifteen years of experience in dealing with male and female inventors.

Is the editor of *Inventors' Digest* magazine an inventor, too? No, but she is a woman.

The following inventions were all created by women: The first washing machine (1871), the first dishwasher (in 1872), the first car heater (1893), the first medical syringe (1899), the first windshield wipers (1903), the first refrigerator (1914) and the first engine muffler (1917), as well as bullet-proof vests, fire escapes, laser printers, flat-bottomed grocery bags, Liquid Paper ®, Scotchgard ®, Kevlar ® and COBOL computer language.

MYTH # 3:
When a person receives a patent on an invention, he is likely to become wealthy from it.

REALITY:
Nothing could be further from the truth, as only about two percent of the 100,000 patents issued by the U.S. patent office each year become profitable. Proof can be found in Ted van Cleave's book, "Totally Absurd Patents," which shows examples of such patents as a pet petter, a Santa Claus detector, a motorized ice cream cone and toilet seat landing lights. When a new invention becomes a hot new product, we hear about it because it's newsworthy. There's nothing newsworthy about the tens of thousands of inventions that fail each year, which is why we never hear about those.

MYTH # 4:
If two people invent the same product at the same time, the one who files for a patent first is awarded the patent.

REALITY:
In the United States, our patent laws are based on a "first to invent" system, while the patent laws in other countries are

based on a "first to file" system. This is why it's important for the inventor to keep good notes of his invention while he's creating it and refining it.

There's a little-known example of a well-known invention that bears this out. In 1876, an American named Elisha Gray filed for a U.S. patent two hours after Alexander Graham Bell filed for his version of the same invention – the telephone. As we all know, Bell won out over Gray, not because he got to the patent office first, but because his notes proved that he conceived of the telephone before Gray did.

So there you have it ... and now you know!

Find out which invention saved former President Bush's life

Lieutenant George Bush, who had become the youngest Navy pilot in history on his eighteenth birthday in 1942, became the only president in history to parachute out of an airplane. This little-known story has an even littler-known controversial twist to it.

Bush parachuted out of the torpedo bomber airplane that he piloted when it was hit by the Japanese and caught fire over the South Pacific during World War II in 1944. When he bailed out of his plane, he pulled the parachute ripcord too quickly and hit the tail of the plane. He was later picked up out of the water by a Navy submarine.

Two of Lt. Bush's crew members died on that fateful flight. One man bailed out but his parachute failed to open, while Bush either thought the other man had already bailed out or was wounded or dead when he didn't answer his intercom. Either way, he went down with the airplane.

Was Bush responsible?

According to the U.S. Veterans Dispatch web site, a man named Chester Mierzejewski, who was approximately 100 feet in front of Bush's plane when Bush parachuted out of it – "so close he could see in the cockpit" of Bush's bomber – claimed that the future president could have avoided losing his crew members crash-landing the plane in the water, rather than parachuting out of it as it began to burn over the Pacific Ocean. Mierzejewski broke his silence about his version of the story in 1997; in Bush's defense, one can only wonder how hard it would be to respond in that type of situation, with events unfolding at lightning speed with no time to spare.

Who invented the parachute?

As is the case with many inventions, there's more than one inventor who contributed to it. Leonardo da Vinci first conceived of the idea in 1483. It's believed that the ancient Chinese also created sketches of what a parachute might look like.

Legend has it that sometime around 1794, Frenchman Jeanne Pierre Francois Blanchard, who was the first person to fly a hot air balloon in America, built the first parachute and tested it using a dog. Blanchard's affair is unsubstantiated and may or may not be true, but it is true that his wife worked as air service chief for Napoleon when he planned to attack England with an invasion of hot air balloons.

A Frenchman named Andre Garnerin made the first parachute jump in 1797. Garnerin would charge fair goers to see him make a parachute jump from his hot air balloon, and then skip town before it was time to jump. One day, before he could escape with their money, the authorities were called and, given a choice of either jumping or going to jail, Garnerin made his first jump, becoming the world's first parachutist. His parachute was based on da Vinci's original design and consisted of a silk pyramid tent with a wicker basket hung from it.

Captain Thomas Baldwin and his brother improved the parachute in 1885. They tested their design by using weighted sandbags from nearby cliffs (rather than their dog) and later used the parachute for the jumps Thomas made from a hot air balloon. Baldwin became known as the "father of the modern parachute," but he and his brother never patented their parachute because they didn't think it would catch on with others. He later designed the first dirigible for the Navy as well as his own airplane, the *Red Devil*.

The arrival of the Wright Brothers' airplane in 1903 expanded the use of parachutes beyond hot-air balloons.

Stefan Banic, a Slovakian inventor who immigrated to Greenville, Pennsylvania, in 1907, invented the parachute which is used in airplanes today. According to Slovakopedia.com, he made a prototype and tested it by jumping from a 41-floor building in 1913 and from an airplane in 1914. He was awarded U.S. Patent # 1,108,484 during the same week that the great war began in 1914 (at the time, World War I was known as "The Great War" because it was the only world war at that point in history). Banic's parachute was used by U.S. pilots during the war.

The impact of the parachute:

In addition to all the lives that the modern parachute has saved since its introduction in 1914, we wouldn't have heard of the men who became our 41st and 43rd presidents, George H.W. Bush and George W. Bush, without it.

Left-handed Leonardo – a clairvoyant and an inventor

Part of this story was inspired by having a left-handed father. I remember his story of how one of the nuns who taught him in grade school made him write right-handed, which was opposite-handed for Dad. Since Mom is right-handed, she was the one who taught us kids how to tie our ties, so that we wouldn't tie them backwards.

The practice of making left-handers write right-handed dates back to at least the mid-1400's when a famous left-handed inventor, Leonardo da Vinci, was temporarily forced to write right-handed by his teachers. I guess Dad could have taught us how to tie our ties "right" by using a mirror, which is what you would need to read da Vinci's notes.

When Leonardo (pronounced lay-uh-nardo) recorded his inventions and discoveries in his notebooks, he wrote backwards, from right to left. Some people believe that he did this in order to prevent people from copying his ideas back in the day before there were patents to protect his inventions, but that explanation is sketchy at best because his writings could easily be de-ciphered with a mirror. His reason for writing backwards is not known.

Da Vinci's inventions:

Let's review a few well-known inventions, along with the name of the inventor who originally designed them, and the year in which they were first conceived:

- Submarine – Cornelius Van Drebbel – in 1620
- Bicycle – 1817 – Baron Karl von Drais of Germany
- Modern scissors – 1893 – Louise Austin
- Flying machine – 1903 – Orville and Wilbur Wright
- Parachute – 1913 – Stefan Banic

Now let's try it again ... with the correct answers for the *original* inventor:

* Submarine – Leonardo da Vinci
* Bicycle – Leonardo da Vinci
* Modern scissors – Leonardo da Vinci
* Flying machine – Leonardo da Vinci
* Parachute – Leonardo da Vinci

All of these inventions were originally conceived hundreds of years earlier by Leonardo da Vinci. Some of his designs were improvements to ideas that had existed earlier, such as the scissors, while his early designs for a flying machine depended on a bird's motions rather than an engine. His final design for a helicopter was eventually shown to Igor Sikorsky, the inventor of the modern helicopter, and it is believed that this drawing inspired Sikorsky to study helicopter design.

Leonardo, one of the Renaissance period's greatest minds, was a painter, architect, musician, sculptor, engineer, scientist and inventor. Born in Vinci, Italy, near Florence, in 1452 as an illegitimate child, he's one of the most proficient left-handed inventors in history, along with Ben Franklin. He painted the famous Last Supper and the Mona Lisa, which I've seen up close at the Louvre in Paris. It was by far the most popular exhibit in the Louvre that day.

Leonardo was a visionary who saw possibilities centuries before others saw them. Some of his inventions required the use of components that had not even been invented yet, such as his design for a rocket that required an engine that would not be invented until four centuries later.

When he died, he left notebooks of his ideas and drawings to a friend and student named Francesco, who later willed them to his son. Francesco's son didn't have any interest in Leonardo's works, and he was careless in preserving them; he would even sell many of them to the first available buyers. For as much as is known about Leonardo da Vinci, there is much more that would be known about him if his works had been better preserved after he died.

Some of Leonardo's sculptures and paintings were later destroyed in subsequent wars and conflicts, which is kind of ironic because he had designed several machines used in war, including machine guns, assault tanks and submarines. He was a peaceful man who detested war. He once designed a tank in which the front and rear wheels moved in opposite directions. Unlikely that a person of Leonardo's abilities would make such an obvious mistake, it makes you wonder whether he designed it with such a flaw so that it wouldn't work, or so that enemies couldn't use it against his people?

Da Vinci, a giant of the Renaissance, a rival of Michangelo and a contemporary of Rafael, was a man ahead of his time. A few hundred years ahead of his time, to be exact.

These celebrities co-starred as real-life inventors

QUESTION:

What do the following famous people have in common?

- Abraham Lincoln
- Zeppo Marx
- Hedy Lamar
- Jamie Lee Curtis
- Michael Jackson

ANSWER: They all received U.S. patents on their inventions. This is probably the only time you'll ever see Abraham Lincoln mentioned in the same story as Michael Jackson.

Now try to match the inventor with their inventions:

- A diaper equipped with a pre-moistened baby wipe
- A method of creating an anti-gravity illusion
- A device for buoying vessels over shoals
- A wristwatch for cardiac patients
- A secret communication system designed to help the allies in World War II

Here's what they invented:

President Lincoln was issued Patent # 6,469 for "A Device for Buoying Vessels Over Shoals" in 1849 while he was still a Congressman in Illinois. It was never commercialized, but a wooden model of the device is on display at the Smithsonian Institution.

Zeppo Marx, whose second wife Barbara later became the fourth wife of Frank Sinatra, was the youngest Marx brother, the one who Groucho said was "off screen, by far the wittiest and funniest of the brothers." Zeppo patented a 1969 wristwatch for cardiac patients. It had two dials; one driven by the pulse of the wearer and the other keeping the steady beat of a normal heartbeat. Zeppo died of lung cancer in 1979 at the age of 78.

Silver screen superstar Hedy Lamarr, born Hedwig Eva Maria Kiesler in 1914 in Vienna, Austria, teamed up with composer George Antheil to patent an invention that manipulated radio frequencies and was intended to prevent the Nazis from intercepting radio-guided torpedoes in World War II. Lamarr personally knew both Hitler and Mussolini when she was married to a pro-Nazi arms dealer, the first of her six future ex-husbands.

The Navy rejected Lamarr's "Secret Communication System" in World War II but the patent, which was issued in 1942 and expired in 1959, served as a foundation in developing technologies that were used in the Cuban Missile Crisis, the Vietnam War and the Gulf War. The U.S. government kept the patent confidential until 1981 because it was under secrecy orders for national security reasons.

The technology, which is similar to what happens when you hit the "scan" button on your car radio, was originally known as "frequency hopping" and is now used in cell phones, pagers, wireless internet devices and defense satellites. Hedy Lamarr died in 2000 at the age of 86.

Jamie Lee Curtis, star of the 2003 hit movie, *Freaky Friday*, along with *Trading Places*, *The Fish that Saved Pittsburgh* and many other films, received U.S. Patent # 4,753,647 in 1988 for a diaper that holds a pre-moistened baby wipe.

Finally, we can't forget the gloved one. Michael Jackson -- yes, *that* Michael Jackson -- is listed on the patent of a "method of creating an anti-gravity illusion" as a co-inventor along with two of his stagehands. This invention allows a person to "lean forward beyond his center of gravity by ... wearing a specially designed pair of shoes." The shoes attach to the stage to allow

Jackson to lean far beyond his center of gravity. In the past, this was accomplished with cables controlled by stagehands. The anti-gravity shoes, on the other hand, allow him to perform the illusion during a live show. The patent issued in 1993.

There are additional celebrity inventors who have patented their inventions, and we'll cover some of them in a future story.

The Story of the Leaning Tower of Pisa ... and the Illinois inventor who figured out how to straighten it

How it all began:

In 1172, a widow named Berta di Bernardo left sixty coins in her will for the purchase of some stones for the creation of the Leaning Tower of Pisa. Back then, it would just be called the Tower of Pisa, since it hadn't begun to tilt until 5 years later. Construction of the Tower began on August 9, 1173 and it's unknown who designed the Tower.

Standing 187 feet tall, the Tower is built on a riverbed only 6 feet above sea level. The ground beneath the Tower is made up of uneven layers of sand and clay, which caused it to lean to the north just 5 years after construction began, while they were working on the third floor. The tilt was then over-corrected, and the Tower began leaning to the south. Subsequent efforts to correct the Tower made it lean even further, leaving the top 17 feet from where it was originally planned. This flaw is what gives the Tower its notoriety and makes it stand out as a tourist attraction.

Construction was interrupted several times, mainly due to wars, and the Tower was finally completed in 1350. It's believed that if construction had been allowed to continue uninterrupted, that the ground would not have had time to settle properly and the Tower would have toppled over.

The story of one man's efforts to straighten it out:

There had been many attempts to correct the tilt over the years, but each one failed. Then Richard Wright, a 77-year old inventor from Collinsville, Illinois, came up with a solution, one that he had been working on since the early 1970's.

In 1995, he sent a letter by certified mail to the mayor of

Pisa, Italy. In the letter, which was signed for and received by
the mayor's office, Rich explained how the Tower could be
straightened with a heavy weight to counteract the tilt. With
the diagrams that he drew up, he showed that they should
counteract the weight at the bedrock level below the ground,
which could be achieved by erecting another building next to
the Tower on the opposing north side. "Weight is weight, and
the new building could be used as a tourist device. The more
tourists that visit the building, the more weight that is used,"
Rich explained. His solution would lead to a gradual
correction, which is what was needed.

Roughly a year later, a TV story showed that Tower officials
had begun to straighten the Tower by using weights to
counteract the tilt. "The method that I suggested was put to
use," Richard said, "but instead of using a building as a
weight, they drilled into the bedrock on the opposite side of the
tilt and added many tons of lead ingots above the ground."

They used Rich's idea, but not his plan. Then something went
wrong, and the Tower over-corrected. No one knows for sure
what happened, but one possible explanation is that they tried
to freeze the foundation under the Tower with liquid nitrogen
and then started removing stones. They didn't realize it at the
time, but these stones were actually part of the Tower's
foundation. They used Rich's idea of counterbalancing the
Tower with weight, but their method was different, and the
result was that the correction was too large.

Maybe they should have used Rich's plan instead.

I've known Rich for six years now. I met him at an inventor's
meeting in 1998, and we soon began working together to get
one of his inventions onto the market. He wasn't looking for
any kind of payment or reward for his Tower efforts. Just
knowing that his solution might straighten the Tower – and
that he might be changing the Tower's course of history – was
enough reward.

Could it have been a coincidence? Sure, but we'll probably never know. Rich continues to invent because it's what he loves to do. Like a lot of inventors, his mind is always working ... trying to solve more problems and develop new products.
The Tower is one of the Seven Wonders of the World. What would happen if they tried to straighten the Tower to make it perfectly level? The people of Pisa would never allow it to happen. After all, the Leaning Tower of Pisa ... must lean.

Did Rube Goldberg ever invent anything worthwhile?

Who was (or is) Rube Goldberg? Is he still alive? Was he a real person or a fictional character?

Most senior citizens and baby boomers think that Rube was an eccentric inventor who created elaborate contraptions to accomplish ordinary, simple tasks. Most of the Generation X group like myself doesn't know whether Rube was real or fictional. Some younger people have never even heard of Rube Goldberg.

Rube Goldberg was a popular cartoonist whose work appeared in newspapers throughout the United States from the early 1900's to the 1960's. His drawings included sports cartoons, comic strips and political cartoons, but he is best known today for the complicated machines that he drew.

Born in San Francisco in 1883, he earned a degree in engineering upon his father's insistence. This engineering background served as a basis for his cartoons of machine contraptions that would take an easy task, such as swatting a fly, and require at least a dozen steps to accomplish it. Rube made sure that every one of the machines in his drawings could work.

Webster's New World Dictionary describes Rube Goldberg as an adjective: "Designating any very complicated invention, machine, scheme, etc. laboriously contrived to perform a seemingly simple operation."

To illustrate this point, take a look at a typical Rube Goldberg invention – his method for a simple fly swatter – without the drawing:

Carbolic acid (A) drips on a string (B) causing it to break and release elastic of bean shooter (C) which projects ball

(D) into bunch of garlic (E) causing it to fall into syrup can
(F) and splash syrup violently against side wall. Fly
(G) buzzes with glee and goes for syrup, his favorite dish.
Butler-dog (H) mistakes hum of fly's wings for door buzzer
and runs to meet visitor, pulling rope (I) which turns
stop-go signal (J) and causes baseball bat (K) to sock fly
who falls to floor unconscious. As fly drops to floor,
pet trout (L) jumps for him, misses, and lands in net
(M). Weight of fish forces shoe (N) down on fallen fly and
puts him out of the running for all time. If fish catches the
fly, the shoe can be used for cracking nuts.

In addition to being a Pulitzer Prize winning cartoonist and an
engineer, Rube was also a sculptor and author, but it was his
cartoons that earned him fame and fortune. Many people think
that the inventor pictured in his cartoons is Goldberg himself,
but it was actually a fictional character that Rube named
Professor Butts.

What do Dilbert®, Calvin & Hobbes®, Garfield® and Bart
Simpson have in common with Snoopy®, Blondie®, Hagar the
Horrible® and Beetle Bailey®? Or, to put it another way, what
do they all have to do with Rube Goldberg?

Their creators are all past winners of the National Cartoonists
Society's Reuben Award, which is given out annually to the
year's top cartoonist. The "Reuben," as you may have guessed,
is named after Rube Goldberg, the Society's first president.

His legacy also includes the various Rube Goldberg Machine
Contests that are held each year among engineering students,
which honor him by designing machines that use the most
complex processes to complete a simple task.

Rube Goldberg succeeded while tens of thousands of other
people who created cartoons, inventions and sculptures failed
to get them off the ground. Rube Goldberg succeeded by taking
an easy task and telling how to devise a complicated
contraption to achieve it.

He died in 1970 at the age of 87. Today, more than years later, his name is synonymous with inventions -- even though he was not an inventor himself.

What do celebrities know about inventing that the rest of us don't?

In one of last month's columns, we profiled the inventions of several celebrities. Today we look at a few more celebrity inventors who applied for and received patents on their inventions.

Harry Houdini received a patent in 1921 for a diver's suit. His diver's suit was meant to allow a deep-sea diver to remove the suit by himself if he was in danger. While Houdini could escape from just about any type of device, he realized that others could not.

Comedian Danny Kaye received a patent in 1952 for a toy that used one mouthpiece to simultaneously unfurl three blow-out paper snakes used at birthday parties.

Steve McQueen was awarded a patent on the bucket seats used in his Ford Mustang in the 1968 movie, "Bullitt."

Actress Julie Newmar, who wore a skin-tight outfit in her role as "Catwoman" in the old *Batman* TV series, patented ultra-sheer, ultra-snug pantyhose. She appeared in the recent movie, *"To Wong Fu, Thanks for Everything, Love Julie Newmar"* and on TV in guest appearances in *Bewitched, The Beverly Hillbillies* and *Star Trek* in the 1960's and most recently in *Melrose Place*.

Celebrity Mom Christie Brinkley created a set of educational blocks for kids, while **Director Steven Spielberg** received a design patent in 1998 for a switch used on mobile camera equipment.

Musicians Eddie Van Halen and **Harry Connick, Jr**. also received patents on their inventions. Van Halen's patent was for a hands-free guitar support, while Connick received Patent # 6,348, 648 last year for his method of displaying written music on computer screens. "It basically eliminates old-fashioned sheet music," said Connick.

What do celebrities know about inventing that the rest of us don't?

Absolutely nothing!

It's not that difficult to get a patent. In fact, it reminds me of the true story of a man who wanted to prove that almost anyone could become a Kentucky Colonel if he had good credentials. So he sent in an application for his dog to become a Colonel and, sure enough, his dog became a Colonel. While getting a patent is not as easy as becoming a Kentucky Colonel, it does require three steps:

Step # 1.
   Create something that is new, useful and non-obvious to the average person in the industry.

Step # 2.
   Conduct a patent search to determine if a similar product has already been patented. If there's no previous patent that would prevent you from obtaining one, then you write the application. It is possible to do this on your own, but most people choose to hire a patent attorney.

Step # 3.
   File the application with the patent office. The average cost, including attorney fees, is around $4,000 for an individual inventor or small company, while the fees for a large corporation are much higher.

Here's an example of what is meant by "non-obvious" in Step # 1: The person who invented the Philips screwdriver was able to get a patent because the second groove – the one that set it apart from a regular screwdriver – wasn't obvious to the average person. But it would now be impossible to get a patent on a screwdriver with three grooves because it would be an obvious difference.

It's common among inventors to create products that relate to their particular area of expertise. You probably noticed that each of our celebrity inventors created products that relate to their craft. And while it's not difficult to get a patent, fewer than two percent of the 6 – million patents that have been issued since the patent office opened in 1790 have produced a profit for the inventor. That's worth considering the next time you come up with a great new idea.

Australia ... Birthplace of boomerangs, sport utility vehicles and black box flight recorders

Welcome to Australia, the land "down under." Australia is the largest island in the world and is home to Aborigines, Tasmanian Devils, Koala bears, kangaroos and the Sydney Opera House. It's also home to some of the world's most unique and valuable inventions.

Australia served as a penal colony for England in the 1800's. In 1894, Australia became the first country to allow women to vote. Called "suffrage," it sounds bad but it's really a good thing. Australia is also the only country to have participated in every modern Olympics.

Our family has been particularly interested in Australia ever since my brother spent two seasons playing pro baseball there in the late 1980's. Many Americans know very little about Australia, and this story focuses on three well-known inventions that originated on this island continent: Boomerangs, sport utility vehicles and the "black box" flight recorders used in airplanes.

1. The Boomerang:

Boomerangs were likely invented for sport by Australian Aborigines a couple thousand years ago. While it's generally believed that boomerangs were used for hunting, it would be nearly impossible to use a boomerang to kill any kind of animal large enough to be worth eating. No one knows exactly what makes a boomerang fly the way it does. The name boomerang comes from a tribe in New South Wales.

2. The Sport Utility Vehicle:

The story of the utility vehicle, or Ute for short, began in 1932 when a farmer's wife asked the Ford Motor Company in

Geelong, Victoria why they couldn't make a vehicle that could haul the family to church on Sunday and the pigs to market on Monday. The company's designer, 22-year old Lewis Brandt, designed a vehicle that combined a truck bed with the cab of a car. It was commercialized two years later.

3. The Black Box:

Dr. David Warren was investigating a series of airplane crashes for the Aeronautical Research Laboratories in Melbourne in 1953 when he invented the "Black Box" flight data recorder. He based his work on the belief that a flight crew might know what went wrong when a plane crashes and that their conversations would provide some clues. The nearly indestructible device, which was originally painted bright red or orange in order to make it easier to locate after a crash, was in production by 1957.

Like many popular inventions, the "black box" was at first deemed unnecessary, by the Australian government, no less. Fortunately, a British company decided to commercialize it, and in 1960 Australia became the first country to make flight recorders mandatory in aircraft, just three years after it first hit the market. Today, nearly every large aircraft in the world has a "black box."

Other notable inventions that come from the land down under include:

- Private ballot box, invented in Victoria in 1856
- Pre-paid postage that is used in nearly every country today, invented by the postmaster general of New South Wales in 1838
- Australian football, better known as rugby, invented in 1858
- Electric drill, invented in Melbourne in 1889
- Speedo swimsuit, invented as "racing swimwear" in 1927, soon followed by the world's first swimsuit competition

- Inflatable airplane escape slide, developed by a Qantas Airlines employee in 1965
- Cochlear ear implant, invented by a professor at The University of Melbourne in 1979

G'day, Mate!

The future looks bright for these four inventions

As one of the judges for a nationwide "New Invention Hunt" that began in August, I had a chance to preview a number of interesting inventions before they're made available to the general public. Now it's your turn for a sneak preview.

Run by the United Inventors Association and co-sponsored by *Inventors' Digest* magazine, Proctor & Gamble and the Academy of Applied Science, the New Invention Hunt had nearly four hundred inventors submit their patented and patent-pending inventions to compete for prize money, prizes and media exposure. Here are four inventions that I thought were especially interesting, and you'll probably see these on the market soon:

The Invisible Writer:

Michael Hall of northern California invented a pen with invisible ink. The ink remains invisible except when viewed under the light source contained in the pen.

Other types of invisible ink become visible only when viewed under an ultraviolet light, or else a chemical must be applied in order to view the ink. The problem with the ink is that it can't become invisible again – it stays visible once the chemical is applied. The Invisible Writer, on the other hand, uses LED's similar to the numbers on your microwave or VCR, and the ink remains invisible except when viewed under the pen's light source. The Invisible Writer is the same size as a regular pen and is much cheaper than other invisible inks which require a chemical.

The Parkinson Glove:

This clever invention comes from 16 year-old Michael Schuman of Ft. Myers, Florida. The Parkinson Glove helps stabilize hand and arm tremors in Parkinson and essential tremor patients. Inspired by an idea his grandfather had told him about, young Mr. Schuman made the Glove out of spare parts from knee and elbow skating guards, and had his mother and grandmother sew it together for him. When he tested it, he noticed a 63% improvement in Parkinson patients and a 58% improvement in essential tremor patients and it won first place in a previous invention contest. He is currently contacting companies about manufacturing it.

The Hydristor:

Thomas Kasmer of Binghampton, New York, invented and patented the "Hydristor," which he claims will double the gas mileage of SUV's. The name of his invention comes from the words "hydraulic" and "transistor." It is currently being tested with a couple of large manufacturers. If all goes according to plan, SUV's will experience a doubling of fuel economy by slowing the engine to an idle at highway cruising speeds, and it will cut the acceleration time in half. Kasmer has two patents on the Hydristor and two more are pending.

Extension Cord Spacesaver:

Can't find an extension cord when you need one? Richard Harper of Mesquite, Nevada, has created a way to store a spring-loaded extension cord into an electrical outlet. He currently has a prototype of the invention and, when it's finished, the cord will be stored inside the wall and will re-coil back into the wall when finished. The Extension Cord Spacesaver is patent pending with a copyright. Now why didn't you think of that!

Who invented television ... an American, a Russian-born immigrant or a Scot?

"This is a beautiful instrument. I wish I had invented it myself."
– Vladimir Zworykin

Many people believe that television was invented by General Electric or RCA (which stands for Radio Corporation of America), but can't remember which big company invented it.

Actually, television was invented by an independent inventor working alone. There were three inventors trying to develop television at the same time:

1. Philo Farnsworth, a 15-year old farmboy from Idaho who rode his horse to school each day
2. Vladimir Zworykin, a Russian immigrant born in 1889 who worked for RCA
3. John Baird, a Scottish inventor born in 1888

Philo Farnsworth, whose grandfather settled with Brigham Young, was born in a log cabin in 1906. According to his ninety-five year old widow, Elma Farnsworth, he decided at age six that he was going to be an inventor when he grew up.

Farnsworth conceived of what television should look like while plowing one of his family's potato fields (although I doubt this is where the term "couch potato" comes from), and he drew illustrations on the chalkboard for his high school chemistry teacher to see.

In his early twenties, he turned down job offers from both RCA and GE, choosing to go it alone. Both of these companies had spent millions of dollars trying to develop television. RCA had also waged a 7-year legal battle with Farnsworth over his patent rights.

A major part of Farnsworth's battle with RCA came from

Vladimir Zworykin, who had developed an electronic method of scanning an image for RCA in 1925. After Zworykin was finally issued his patent thirteen years later, he couldn't produce any evidence to prove that he had constructed and operated his system before Farnsworth did, and RCA lost the case.

Across the ocean, there was another inventor obsessed with inventing the first working television. John Baird sent what he called 'pictures by wireless' in 1923, and then sent and received the first wireless television signal two years later. In 1928, he became the first person to broadcast live images across the Atlantic and he started broadcasting with the BBC regularly in 1929. But the process with which he did all this – known as "mechanical scanning" -- soon became obsolete.

Despite the competition with John Baird and the financial backing that RCA provided to Vladimir Zworykin, it was Philo Farnsworth who became the father of television. So did Farnsworth live happily ever after?

Unfortunately, no. After beating Zworykin and RCA in court, Farnsworth was paid a handsome royalty for the right to license his television, which marked the first time RCA paid a royalty to anyone. Even though he developed modern television, RCA brought it to market first and began regular broadcasts in 1939 through NBC, which it owned.

By 1941, Farnsworth was ready to follow RCA onto the market, but the United States government soon banned commercial television during World War II. By the time the war ended, Farnsworth's patent had run out, and so did his luck. While he profited from the licenses that he sold, those licenses ran out when his patent expired.

More than a decade after his death in 1971, Farnsworth finally received some of the credit that he deserved. The U.S. Postal Service commemorated him with a stamp in 1983, and he was given an honorary television Emmy Award in 2001.

Time magazine recognized him as one of their "100 Most Influential Persons of the 20th Century." By 1951, there were ten million TV sets in the United States and it is estimated that there are now more people who own a TV set than a telephone.

A hunter's dream: A story about fishing reels, bows and arrows and guns

"Give a man a fish and he has food for a day; teach him how to fish and you can get rid of him for the entire weekend."
– Zenna Schaffer

As hunting season approaches, we look at the origin of the hunter's toys: The bow and arrow and the rifle. Since I've never known a hunter who didn't also fish, we explore the origins of fishing, too.

The origins of fishing ...

Fishing began in the Stone Age (and a caveman called in sick the following Monday to go fishing). Before there were fishhooks and fishing string, pre-historic fishermen used spears to catch fish. Later, the more "modern" pre-historic fishermen used a gorge to catch fish. A gorge was a baited piece of bone or flint with two sharp ends and a leather line attached at the middle.

Angling, which is fishing for sport rather than for food, dates all the way back to the Old Testament. It's not known when the first basic fishing pole (without a reel) was used, but one source indicates that the Egyptians fished with rods, lines and hooks as early as 2000 B.C. The first drawing of a fishing pole was from the Orient in 1195.

In the 1650's, England's legendary Charles Kirby developed the bent hook that we use today. In 1820, George Snyder of Paris, Kentucky, became the first American to produce fishing poles with reels. Originally, the reel was used mainly for storing excess string. It's possible that the British made fishing poles with reels around the same time as Snyder on "the other side of the Pond," although there are no records to verify or dispute this.

Hunting with bows and arrows ...

Pre-historic hunters used bows and arrows more than 8,000 years ago, although it's possible that the bow and arrow dates as far back as 25,000 B.C., which would be about the same time as the first boomerang. Contrary to popular opinion, the boomerang was not effective for hunting.

The bow and arrow wasn't the first hunting tool, though, as stone axes and spears preceded it. The crossbow was invented in the Middle Ages around the late 1500's, and its silent nature benefited hunters then as it does today. Even in 2003, there are still primitive tribes of people who use bows and arrows to hunt down their meals.

From muzzle loaders and muskets to the 21st century ...

Gunpowder was invented in China around 1040 for use in fireworks and rockets; it wasn't until after it arrived in Europe a couple of centuries later that it was first used in guns.

The first known reference to a gun was in 1326, although it was called a vaso because it resembled a vase. It bore no resemblance to any modern guns; in fact, it was fired like a gun but it shot an arrow rather than a bullet. Since it was called a vaso, it probably came from Italy. Early guns were fired with burning sticks or hot coals.

Shotguns were used to hunt small game as early as 1549. Single-barrel shotguns were followed by double-barrel shotguns in the late 1700's. Early muskets were muzzle-loading guns which were set off with a lighted match. Muskets first appeared in the 1600's and were replaced by rifles around 1850.

Whether you hunt with a shotgun or a .22 – a Remington, a Ruger or a Winchester – chances are that your gun is similar to those made a hundred years ago.

My time's up. It's time to take my nephew snipe hunting.

Who really invented baseball ... Alexander Cartwright or Abner Doubleday?

"...And somewhere men are laughing, and little children shout;
But there is no joy in Mudville – mighty Casey has struck out."
 – Ernest Thayer, from the poem "Casey at the bat,"
   June 3, 1888

Major League Baseball celebrated the 100th anniversary of the World Series in October of 2003. In the first-ever World Series in 1903, the Boston Pilgrims (Red Sox) defeated the Pittsburgh Pirates, 5 games to 3. The Series was originally a best-of-nine format.

The 2003 season also marked the 100th anniversary of the event that started the great debate over who "invented" baseball. In this story, we try to find out who invented baseball. There are two competing stories, and they involve two men who were born within a year of each other and died within a year of each other. In fact, both men had died by the time the great debate began. It was either bank clerk Alexander Cartwright or Civil War veteran Abner Doubleday, whose great-great-grand-nephew is the co-owner of the New York Mets.

How the Debate Began ...

The debate began when baseball writer / historian Henry Chadwick, who wrote baseball's first rulebook in 1858, declared in Albert Spalding's *Baseball Guide* of 1903 that baseball had been derived from an English game called "rounders."

Al Spalding was a former major league pitcher and manager for the Chicago Cubs (originally known as the Chicago White Stockings). Since he didn't want to accept that the game he loved could have come from the British, he commissioned a panel in 1904 to determine the game's origins. The panel, which included two U.S. senators and was chaired by a former National League president who probably never heard of Alexander Cartwright, also didn't want to accept the possibility that baseball might have British roots. Their choice as the inventor of baseball was a Civil War general named Abner Doubleday. Doubleday, by the way, has the distinction of being the soldier who fired the first shot in defense for the Union during the Civil War, at Fort Sumter.

The only evidence that the panel had in support of Doubleday was a letter they received from an elderly man who claimed that he was a boyhood friend of Doubleday's. In his letter, he claimed that he saw Doubleday invent baseball in Cooperstown in 1939 when he organized two teams in a game which included bases and a ball. Most of the other research for this panel was done by an employee of the publishing company which Spalding owned.

There was plenty of evidence to suggest that Doubleday did not invent baseball, though. For example, Doubleday kept diaries and was a skilled public speaker, but there was never any mention of baseball in his writings or his speeches. You would think that a person who invents a new sport would mention it somewhere along the way.

Alexander Cartwright, on the other hand, established many of baseball's basic rules. He established that the distance between bases is to be 90 feet, that the game is to be played by nine-person teams for nine innings, and that each team gets three outs per inning. In addition to adding the position of shortstop, he eliminated the rule that allowed the defense to get a runner out by throwing the ball at him! He also divided the field into fair and foul territory. Many believe that September of 1845 is when Cartwright invented the game at age 25, and his Knickerbocker baseball club played their first game the following year in Hoboken, New Jersey.

To further complicate matters, there were claims that there

was a second man named Abner Doubleday, and the game that Doubleday's childhood friend had claimed to see him invent was actually a form of the British-based rounders game mentioned earlier, called "Town Ball." Years later, a baseball with the cover nearly completely torn off was found in this man's attic; it became known as the "Doubleday" baseball and it sits in the Hall of Fame.

Which man is in the Hall of Fame?

Where can you find most of this information about Cartwright's contributions to the rules?

On his Hall of Fame plaque, which also lists him as the "Father of modern baseball." Cartwright's plaque doesn't claim that he invented the game, but he is in the Hall of Fame, while Doubleday is not.

So who did invent baseball – Alexander Cartwright or Abner Doubleday?

You have to decide for yourself. Even though the evidence favors Cartwright over Doubleday, no one knows for sure because there wasn't enough proof at the time – more than 150 years ago. Plus, there were accounts of "baseball" being played as early as the 1820's and 1830's in the Northeast, although those games may or may not have resembled today's game.

Personally, I believe that Al Spalding – whose company, named Spalding, manufactures sports equipment – established his panel for one purpose only – to manufacture an American origin for baseball.

Eureka! Who were Archimedes, Ctesibius and Hero?

Who were Archimedes, Ctesibius and Hero and what did they invent? And are their inventions still being used today, more than 2,000 years later?

Born in Syracuse in 287 B.C. and educated in Alexandria, Archimedes is known as the man who jumped up out of his bathtub one day and ran naked through town shouting "Eureka! Eureka!" In case your knowledge of Greek is as limited as mine, "Eureka" means "I have found it."

Why did Archimedes do this? And what did he find?

While taking a bath, he had solved the dilemma of water displacement; namely, the relationship between the weight and volume of an object in water vs. the weight and volume that was displaced when he got out of the tub.

Other than the term "Eureka," what did Archimedes invent?

- The hydraulic screw, also known as the Archimedes screw, which was used in pumping water from the Nile River
- The worm gear, which is still used today
- A system of ropes and pulleys which he used to move a ship while it was docked on land, effectively creating the world's first winch.

Archimedes is also credited with inventing the world's first catapult, and legend has it that he showed how to use a mirror to focus the sun's rays on an enemy ship, causing it to burn. Known more as a mathematician than an inventor, he also calculated the exact value of pi.

One of his contemporaries, Ctesibius (pronounced ti-sib-e-us) lived in Alexandria around the same time as Archimedes, but the two geniuses probably never met.

Ctesibius invented the water clock, which was known back then as the Clepsydra. The sundial had already been invented but would only work during daylight hours on sunny days. Ctesibius also created three inventions in conjunction with each other:

- The valve, which led him to create his next great invention ...
- The suction-pump, which was used for fighting fires and led to his next great invention ...
- The pump that was used as a source of wind for the first organ.

The most accomplished of the Greek inventors was Hero who, like Ctesibius, was from Alexandria. Hero learned a lot from Ctesibius, but since most of the records of their time have been destroyed, it's not known if Hero lived during the same time as Ctesibius or if he came after him and merely learned from his writings.

Like Italy's Da Vinci, Hero is regarded by history as a man whose work was hundreds of years ahead of his time. Hero created the following inventions:

- The world's first steam engine, which was called an aeolipile. The principle behind the aeolipile was that every action has an "equal and opposite reaction," which we all learned in school. Sir Isaac Newton discovered this 1,600 years later, and makers of jet engines used this same principle 1,900 years later.
- A machine which would dispense a fixed amount of holy water when a coin was put into it. This was the world's first automatic vending machine.
- The screw-press, which extracted olive oil from olives and juice from grapes.
- The odometer, which measured the distance that taxis traveled. He did this by making a pointer with gears that counted the number of revolutions of the taxi cart's wheel. Hero called his invention the hodometer. Ben Franklin

would later invent an odometer to measure the distance that mail carriers would travel for each delivery.

Since no story about the ancient Greeks would be complete without some sort of tragedy, we end this story with the account of Archimedes' death in 212 B.C. When the Romans invaded Syracuse, the Roman ruler ordered that Archimedes be left alone. One of the soldiers didn't recognize him, though, and killed him with his sword.

And that's the end of this Greek tragedy. It's time to go fix myself a Hero sandwich.

What's so interesting about a postage stamp?

"Of all the wild schemes I have ever heard of, this is the most extraordinary!"
   – British Postmaster General in 1840 reacting to the idea of using postage stamps for mail delivery

When I was in college, I took some time off and backpacked throughout Europe. For one month I was a typical college bum, and it was fun. When I went, I decided to bring back a souvenir from each country that I visited – something that was small enough to carry in my backpack and could be found in every country. Quick – what does each country have that would make a good souvenir and is small enough to fit into a backpack?

Stamps! I decided that stamps would be the perfect souvenir.

So I chose coins instead. Since every country requires you to use their currency when you pass through their borders, I knew it would be easier to collect coins rather than stamps. Stamps, though, have a history all their own; they tell a story, just like a country music song does.

The very first postal services were set up by kings and governments exclusively for their own use. Later, when ordinary citizens wanted to send mail as well, a system was established which required the person who received the letter to pay for it at the time of delivery. They were charged according to how much the letter weighed as well as the distance that it went. In fact, Ben Franklin invented an odometer to measure the distance the letter carriers traveled.

Things began to change in 1838 when James Raymond, the Postmaster General of New South Wales, Australia, introduced the world's first pre-paid postage system by stamping letters. It was set up the same way in which a bank teller stamps your checks.

Two years later, an Englishman named Sir Rowland Hill came up with the idea of using postage stamps. Hill suggested lowering the cost of postage to a penny and, since the stamp was black, it was called the Penny Black. The Penny Black contained an image of Britain's Queen Victoria and was first issued in England in May of 1840. The British postmaster general thought that the postage stamp was a crazy idea at the time. Hill's next great idea was the mailbox, now that postage was being pre-paid by the sender. (That Hill thinks of everything, doesn't he?)

Stamps made their way to America in 1847, and Ben Franklin was the first person to appear on a U.S. stamp; he was also our first postmaster general. The five-cent Franklin stamp was soon followed by the ten-cent George Washington stamp.

In 1860 the Pony Express opened with a recruitment ad that read, "WANTED: *Young skinny wiry fellows not over 18. Must be willing to risk death daily. Orphans preferred. Wages $25 a week.*"

The Pony Express riders could travel the 2,000 miles from St. Joseph, Missouri, to the West Coast in only ten days, which was half the time that it took to travel the distance by train. One of the riders was 14-year old William Cody – as in Buffalo Bill Cody – who once outran a party of 15 Indians who were trying to rob him. Cody and his fellow riders delivered news of the outbreak of the Civil War the following year. After just nineteen months, the Pony Express was replaced by the telegraph.

In case you've always wondered, but were afraid to ask ...

- Even though the English invented the postage stamp, they remain the only country in the world that does not put their country's name on their stamps.

- The one-penny stamp from New South Wales, Australia, which showed the seal of the colony, is worth around $5,000 in mint condition today.
- The first person other than royalty to appear on a British stamp was William Shakespeare in 1964.
- The best-selling U.S. commemorative stamp of all time is the 1993 Elvis Presley stamp, of which 124 million have been sold.
- In 1973 the country of Bhutan issued a stamp that looked like a record and would actually play the Bhutanese national anthem.
- Cats were used for mail service in Belgium in 1879, but this experiment failed because the cats weren't disciplined enough to deliver the mail!

And that's a good one with which to end this story.

Will the real inventor please stand up?

The names of Alexander Graham Bell, Charles Darrow and Guglielmo Marconi are synonymous with their inventions. But are they really the true inventors, or could there be more than one inventor?

I've never been one to support conspiracy theories, but many inventions have had more than one inventor working on them at the same time. So while you might not recognize the names of Elisha Gray, Lizzie Magie or Nikola Tesla, you certainly know the inventions for which they may – or may not – have created.

Who's on the phone – Bell or Gray?

While it is Alexander Graham Bell who is credited with inventing the telephone in 1876, there was another person who tried to patent the telephone during the same year as Bell. In fact, Elisha Gray arrived at the patent office to apply for a patent for his version of the telephone on *the very same day* as Bell – just 2 hours later!

Bell was awarded the patent, but the case went to court; in fact, it went all the way to the U.S. Supreme Court. Since Bell kept better records of his design than Gray did, his patent was sustained and the rest, as they say, is history. Elisha Gray did go on to invent and patent the facsimile telegraph system in 1888, while Bell went back to working with deaf children after giving *all* of his American Bell Telephone Company stock to his new bride on their wedding day.

Other inventors later staked a claim to inventing the telephone. Bell's patent, which to this day remains the most valuable patent in history, faced more than six hundred lawsuits. None changed Bell's status as the official inventor of the telephone, though.

Did Charles Darrow have a monolopy on the world's best-selling board game?

Not many people recognize the name of Lizzie Magie as the inventor of the board game *Monopoly*, but then again, not too many people recognize the name of the person who is widely considered to be the inventor of the game, Charles Darrow. In 1904 Lizzie created *"The Landlord's Game"* to teach people the unfairness of realty and tax systems. Soon people were customizing the game to reflect their own neighborhoods.

Charles Darrow of Germantown, Pennsylvania, played one of these games at a friend's house. He then began manufacturing the games himself and selling them for $4 apiece. When he couldn't keep up with demand, he wrote to Parker Brothers to see if they would license it from him. The company rejected him at first, citing 52 fundamental flaws with the game. When they heard how well the game sold during the Christmas season of 1934, they reconsidered. More than half a billion people have played *Monopoly* since.

"Marconi is a good fellow ... he is using seventeen of my patents."

While many inventors contributed to the development of radio, Nikola Tesla is probably its main inventor. Guglielmo Marconi, though, is the one who received credit – and wealth – for the invention of radio.

Tesla filed his basic radio patent applications in 1897, three years before Marconi filed his; as a result, Marconi's applications were turned down. Yet it was Marconi who was the first to transmit and receive signals across the Atlantic Ocean when he signaled the letter "S" from Cornwall, England to Newfoundland, Canada, in 1901.

When one of Tesla's engineers said, "Looks as if Marconi got the jump on you," Tesla replied, "Marconi is a good fellow. Let him continue. He is using seventeen of my patents." But in

1904, for some unknown reason, the patent office suddenly reversed its rulings and awarded the main patent to Marconi. The history books began to refer to Marconi as "the father of radio" when he won the Nobel Prize in 1909, prompting Tesla to sue the Marconi Company for infringement. Tesla lost because he didn't have the funds to finance the case. The patent office reversed its decision in 1943 when the Marconi Company sued the U. S. government for use of its patents in World War I. When it was all said and done, though, Marconi had become a wealthy man while Tesla had gone broke.

What did these inventors do for a living before they became household names?

There's a TV show called *"Where are they now?"* which focuses on the current lives of former celebrities. We take the opposite approach in this article by revealing the jobs that five inventors held *before* they became famous. Some of their backgrounds make perfect sense as inventors, while others may surprise you. We also include the story of one person who didn't invent anything, yet his name is synonymous with a certain type of invention.

LEVI STRAUSS (1829 – 1902)
Twenty-four year-old Levi Strauss left New York for San Francisco in 1853 to open a dry goods store with his sister and brother-in-law. They sold supplies to miners and other products to the people of San Francisco during the Gold Rush days. One of his customers had a method of making jeans with metal rivets and, unable to afford the cost of a patent, he asked Strauss to pay for the patent and go into business together. In May of 1873, the first official blue jeans were made. I think you know how that turned out.

CLARENCE BIRDSEYE (1886-1956)
The name of Birdseye is synonymous with frozen foods, yet many people do not know that there was a person named Birdseye behind it all. Clarence Birdseye's job prior to becoming an inventor is what led him to become an inventor. As a biology major in college, he went to work as a naturalist for the U.S. government and was assigned to the Arctic. There he observed first-hand the ways of the Eskimos who lived there. Birdseye saw how the combination of ice, wind and temperature froze the fish that had just been caught. He also noticed that the fish retained most of their taste when they were cooked and eaten. When he returned home to New York in 1924, he founded Birdseye Seafoods, Inc.

KING CAMP GILLETTE (1855 – 1932)

The work of Gillette's parents laid the groundwork for him to become an inventor. For a while, his father worked as a patent agent and part-time tinkerer, and his mother created a cookbook in 1887 that remained in print for 100 years. Gillette became a traveling salesman at age 17, and he often made improvements to the products that he sold. He learned the importance that disposable items had on sales, and used this concept for his idea for improving the safety razor blade. Production began in 1903 and 100 years later, the company that bears his name rings up nearly $10 billion a year in sales in more than 200 countries. Despite his first name and the success that he had, King Gillette opposed capitalism, and he wrote books in which he declared competition to be the root of all evil.

RON POPEIL (born 1935; still inventing)

Born in 1935, Ron Popeil is famous for demonstrating his inventions on TV. His product line includes the Ronco Spray Gun, Dial-O-Matic, Veg-O-Matic, Mince-O-Matic, Popeil Pasta Maker, Pocket Fisherman and the Showtime Rotisserie Oven. What did this master pitchman do before he began selling his inventions on TV? He pitched his Dad's inventions on the streets of Chicago in the 1950's, as it was his Dad who taught him the basics of salesmanship and showmanship. Popeil's inventions have rung up more than $2 billion in sales, and counting.

RUBE GOLDBERG (1883-1970)

The term "Rube Goldberg invention" has led millions of Americans to believe that Goldberg was an inventor. You won't find his name on any patents or store shelves, though, because ol' Rube never invented anything. After graduating with an engineering degree, he worked as an engineer for a short time but hated the job, so he began doing what he loved most – drawing. Goldberg won a Pulitzer Prize in 1948 for his cartoons depicting elaborate schemes which took 10 or more steps to accomplish a simple task. Goldberg is probably the only "inventor" to be honored with both a postage stamp and an adjective named for him, as in "our Rube Goldberg tax system."

Who's the real McCoy -- an inventor, a boxer or someone else?

There are a number of possible answers to this question, as each McCoy mentioned in this story achieved some notoriety during his lifetime. Another relevant question is, "Does the term refer to some*thing* or to some*one*?"

The contestants are ...

•   Elijah McCoy, a Canadian inventor
•   Kid McCoy, a former welterweight boxing champion in 1896
•   Other
•   None – or all – of the above

Or was the "real McCoy" part of the "Hatfields & McCoys" feud in Appalachia that ended more than 100 years ago?

Inventor Elijah McCoy was born in Ontario, Canada, in 1844 to former slaves who had fled from Kentucky before the Civil War. He was educated in Scotland as a mechanical engineer, and then moved to Michigan. Unable to find a job as an engineer, he went to work for a local railroad company as an oilman.

McCoy's job involved walking the length of the train to oil its moving parts, such as the axles and bearings. Believing that there must be a better way to accomplish this, he invented a lubricating cup that automatically dropped oil onto moving parts. His automatic oil cup was requested by engineers and inspectors until it eventually became standard equipment. It also became known as "the real McCoy" along the way. Elijah McCoy earned a total of 57 patents in his lifetime, and established the McCoy Manufacturing Company in Detroit.

Contestant # 2 was boxer Kid McCoy, whose real name was Norman Selby. Born in 1873, he began billing himself as

Kid "The Real" McCoy in the 1890's after a number of
imposters claimed to be Kid McCoy in order to capitalize on
his fame and fortune. Kid McCoy was a colorful character who
always carried a roll of money with him. He was married
10 times; four of those marriages were to the same woman.

Then there's the story about a boxing match in which McCoy
fought a deaf fighter. It was during the match that he found
out his opponent was deaf, and he offered his "help" by
signaling that the bell had rung during the third round, when
in fact it didn't. As his opponent thanked him and turned
toward his corner, McCoy knocked him out.

In a match against a barefoot fighter in South Africa, McCoy
threw tacks into the boxing ring. He won that bout, too, but he
lost in court after shooting his girlfriend. He served time in
San Quentin State Prison and was later paroled. After his
boxing career ended, he went to Hollywood and acted in a
few movies. He died in 1940.

There are additional versions of who the real McCoy was. The
Oxford English Dictionary quoted Robert Louis Stevenson in
1883 as referring to someone as "The real Mackay," although
it's not clear to whom Stevenson was referring. There was also
an 1880's brand of whiskey that was advertised as
"the real McKay."

Both references occurred a decade *after* Elijah invented his
self-oiling device for locomotives, but *before* Kid became
famous.

Then there was prohibition-era smuggler Billy McCoy, who
imported genuine whiskey into the United States from
Canada. Since his whiskey was real rather than the stuff
made by moonshiners, it was known as the real McCoy. There
was also a cattle baron who Alistair Cooke believed was the
real McCoy. He may have been referring to Charles Goodnight
of Texas and the design of his 1866 chuck wagon which was named
after him (using the common nickname of "Chuck" for "Charles").

For the real "real McCoy," whoever he is, his name became a noun, just like Rube Goldberg's name became an adjective. Yet to this day, nobody knows for sure which person – or product – the term refers to.

It's possible – although unlikely – that "the real McCoy" could have described more than one person – and that each one was given the nickname independently of each other. Some mysteries are better left unsolved, because it keeps the legend alive and growing – and interesting.

Can this inventor prove that the Loch Ness Monster does – or doesn't – exist?

"Why would you want to go to Loch Ness? It's just a big lagoon."
– skeptical British locals in 1988

I have a business partner named Joanne Hayes-Rines. She is the editor and publisher of the magazine for inventors, *Inventors' Digest*. Over the course of our 4-year working relationship, she has told me bits and pieces of her husband's work as a pioneering inventor.

But it wasn't until after I saw a segment on *The History Channel* showing a 1972 interview with Dr. Robert Rines that I decided to find out more about this interesting inventor. While serving as a lieutenant in Saipan during World War II, Dr. Rines invented imaging radar in order to give the soldiers notice of incoming enemy aircraft so they could defend themselves.

After the war, he received his law degree and later completed a Ph.D thesis in China, where he helped improve their patent system. His early work with sonar included bringing definition to it, and his discoveries ultimately led to the development of the modern sonogram as well as the technology that was used in finding the TITANIC and the BISMARCK. Dr. Rines was inducted into the National Inventors Hall of Fame in 1994, and today he holds more than 60 patents.

What does all this have to do with the title of this story – something about the Loch Ness Monster?

The History Channel interview with Dr. Rines mentioned that he is one of the world's foremost experts on the Loch Ness Monster. On June 23, 1972, Dr. Rines saw the back of the monster for about 15 minutes. Later that summer, he and his team captured the image of a flipper with an underwater still camera that snapped pictures every 45 seconds.

When I backpacked around Europe in the Fall of 1988, I had only enough time and money to do one of two things before leaving England ... either visit Stonehenge or Loch Ness. When I asked some of the locals for suggestions, half of them remarked, "Why would you want to go to Stonehenge? It's just a bunch of big rocks," while the other half said, "Why would you want to go to Loch Ness? It's just a big lagoon." Armed with such valuable insights from the locals, I decided that my time and few remaining tourist dollars would be better spent at Stonehenge rather than searching for Nessie. I knew I could count on seeing a bunch of big rocks.

The Nessie.co.uk web site says that Sir Peter Scott, who served as chairman of the World Wildlife Fund, claimed that a combination of underwater pictures and earlier film records convinced him that large animals exist in Loch Ness. He gave them the scientific name of "*Nessiteras rhombopteryx*" in order for them to be protected under British laws. Translated, the name means, "The wonder of Ness with the diamond shaped fin," which was a reference to the underwater photograph that Dr. Rines and his team took in 1972. Later it was discovered by doubters that the letters in "*Nessiteras rhombopteryx*" can be re-arranged to spell "Monster hoax by Sir Peter S," but Dr. Rines countered that the letters can also be re-arranged to spell "Yes, both pix are Monsters R."

We might never know if Nessie has ever existed, but it probably won't be proven that the giant monster has never existed. The legend of the Loch Ness Monster will always have skeptics unless the actual monster is captured or the lake is drained.

So how does Dr. Rines respond to the skeptics who doubt his claim?

He just smiles.

If you enjoyed this book, then you'd probably enjoy reading a new INVENTION MYSTERIES article in your local newspaper each week.

How?

The INVENTION MYSTERIES book is a collection of articles that ran as a weekly syndicated newspaper column in 2003.

Call or e-mail your local newspaper and ask the editor to carry INVENTION MYSTERIES as a weekly column. Editors listen to their readers' suggestions.

If your paper decides not to run the INVENTION MYSTERIES column, then you can still purchase the articles of 2004 in next year's book (volume 2). More details are available at www.InventionMysteries.com

Index

C

E

F

J

Jackson, Michael 21, 22, 47, 86, 87,104, 105, 106
Jefferson, Thomas 11, 12, 13, 67, 68, 69, 86, 87
Jet 21, 133
Jet engine 133
jet-powered surfboard 21
Johnson, Lyndon 55
Johnston, Philip 26
Jones, Davy 37
Julius Sextus Frontinus 17

K

kangaroos 116
Kasmer, Thomas 120
Kaye, Danny 113
Kearns, Robert 58, 59, 60
Kentucky Colonel 114
Kettering, Charles 19
Kevlar ® 55, 56, 96
Kiesler, Hedwig Eva Maria 105
King James I, II, III, IV, VI 83
Kitty Hawk 61
Knickerbocker 129
Koala bear 116
Kwolek, Stephanie 55, 56

L

ladder which enables spiders to climb out of the bathtub 21
lagoon 147, 148
Lamar, Hedy 104
laser printers 42
Last Supper 102
Leaning Tower of Pisa 6, 107, 109
leash for walking an imaginary dog 20
LED's 119
Lemelson Center 56

M

N

R

S

U.S.S. Hopper 43
ultra-sheer, ultra-snug pantyhose 113
United Inventors Association 119
United States 2, 12, 17, 25, 47, 49, 61, 64, 67, 83, 93, 96, 110, 123, 124, 145
Ute 116

V

valve 133
Van Cleave, Ted 96
Van Drebbe, Cornelius l 101
Van Halen, Eddie 133
vaso 126
Veg-O-Matic 142
Volvo 59
von Drais, Baron Karl 88, 101
vote recorder 74, 75

W

waffles 69
Warner Brothers Studios 18, 28
Warner, Harry 18
Warren, Dr. David 117
washing machine 96
Washington, President 11, 12, 13, 46, 62, 68, 136
water clock 133
welterweight boxing champion 144
wheel cipher 69
winch 132
Winchester 126
windshield wipers 42, 58, 59, 60, 85, 96
windshield wipers, intermittent 58, 59, 60, 157
Windtalkers 25
World Series 88, 128
World War II 5, 24, 25, 33, 64, 85, 93, 98, 104, 105, 123, 147
World Wide Web 5, 14, 15, 43, 47
worm gear 132
Wright Brothers 61, 62, 63, 88, 89, 90, 99
Wright, Orville 61, 62, 63, 101

Have you ever invented anything?

If you do, then here are 3 great places to turn to for help:

- Inventors' Digest magazine: The "magazine for idea people." www.InventorsDigest.com

- United Inventors Association of the USA: The UIA is a non-profit corporation formed in 1990 to provide leadership, support, and services to inventor support groups and independent inventors. www.uiausa.org

- MarketLaunchers.com: The "yellow pages of inventions," MarketLaunchers.com builds web pages for inventors and also offers a free online newsletter. Run by Paul Niemann, author of this book. www.MarketLaunchers.com

*"Inventions have long since reached
their limit, and I see no hope for
further developments."*
Roman Engineer Julius Sextus Frontinus,
10 A.D.

*"Everything that can be invented--
has already been invented,"*
former patent
commissioner, 1899

"That's an amazing invention, but who would ever want to use one of them?"
President Rutherford B. Hayes in 1876, after Alexander Graham Bell demonstrated the telephone to him at the White House

*"Heavier-than-air flying machines
are impossible,"*
Lord Kelvin, President Royal Society, 1895

INVENTION MYSTERIES

The little-known stories behind well-known inventions

Paul Niemann

Horsefeathers Publishing   Quincy, Illinois

Easy Order Form:

To order online: www.InventionMysteries.com
To order E-mail: niemann7@aol.com
To order by phone: (800) 337-5758. Please have your
                           credit card ready.

To order by mail: Invention Mysteries
Quincy, IL 62305   P.O. Box 5148
                           Quincy, IL 62305

Name: _____

Address: _____

City: _____ State: _____

Telephone: _____

E-mail: _____

Please check one: 1 book @ $12.95: _____
                        2 books @ 25.90: _____
                        4 books @ 38.85: _____ (buy 3; get 1 free)

Sales tax:
Illinois residents please add 7.75% tax before adding the shipping charge.

Shipping:
$2.95 per order, regardless of how many books you order.

Payment: (Please check one)
Check _____ Credit card _____
Visa   Master Card Discover   AMEX
Card number: _____
Name on card: _____
Exp. Date: ___/___

INVENTION MYSTERIES

The little-known stories behind well-known inventions

Paul Niemann

Horsefeathers Publishing   Quincy, Illinois

Easy Order Form:

To order online: www.InventionMysteries.com
To order E-mail: niemann7@aol.com
To order by phone: (800) 337-5758. Please have your
credit card ready.

To order by mail: Invention Mysteries
Quincy, IL 62305   P.O. Box 5148
Quincy, IL 62305

Name: _____

Address: _____

City: _____ State: _____

Telephone: _____

E-mail: _____

Please check one: 1 book @ $12.95: _____
2 books @ 25.90: _____
4 books @ 38.85: _____ (buy 3; get 1 free)

Sales tax:
Illinois residents please add 7.75% tax before adding the shipping charge.

Shipping:
$2.95 per order, regardless of how many books you order.

Payment: (Please check one)
Check   _____   Credit card   _____
Visa   Master Card   Discover   AMEX
Card number: _____
Name on card: _____
Exp. Date: ___/___